清华大学建筑 规划 景观设计教学丛书
Selected Works of Design Studios: Architecture, Urban Planning, Landscape
Tsinghua University

Atelier International d'Architecture Construite

# 国际建筑工坊

程晓青 邹欢 栗德祥 编著

清华大学出版社
北京

## 内容简介

"国际建筑工坊"是清华大学建筑学专业研究生设计专题的选题之一,源于2002年起由中、法、意、韩等国发起的国际学生建筑设计竞赛,旨在搭建东西方建筑教育交流的舞台。经过十四年的共同建设,已经形成跨越欧洲、亚洲和美洲的国际合作格局。本课程密切关注建筑学科发展前沿和各国城乡建设前沿的关键问题,强调设计选题的研究性,希望增强学生对于不同文化的学习与思考,提高其解决复杂城市与建筑问题的能力。本书撷取清华大学历年参与本课程的部分学生之优秀设计作品,希望向读者展现在不同文化环境下的建筑探索。

本书适合具备一定建筑与城市设计基础的研究生或本科高年级学生,可帮助其拓展专业视角,加深对于异质文化环境和城市建设的认识与思考。

版权所有,侵权必究。侵权举报电话:010-62782989 13701121933

图书在版编目(CIP)数据

国际建筑工坊 / 程晓青,邹欢,栗德祥编著. —— 北京:清华大学出版社,2016
 (清华大学建筑 规划 景观设计教学丛书)
 ISBN 978-7-302-42068-2

Ⅰ. ①国… Ⅱ. ①程… ②邹… ③栗… Ⅲ. ①建筑设计-作品集-中国-现代 Ⅳ. ①TU206

中国版本图书馆CIP数据核字(2015)第264029号

责任编辑:周莉桦
封面设计:张华西
责任校对:王淑云
责任印制:宋 林

出版发行:清华大学出版社
 网　　址:http://www.tup.com.cn, http://www.wqbook.com
 地　　址:北京清华大学学研大厦A座　邮　编:100084
 社 总 机:010-62770175　邮　购:010-62786544
 投稿与读者服务:010-62776969, c-service@tup.tsinghua.edu.cn
 质量反馈:010-62772015, zhiliang@tup.tsinghua.edu.cn

印 装 者:北京亿浓世纪彩色印刷有限公司
经　　销:全国新华书店
开　　本:165mm×230mm　印　张:12.5　字　数:328千字
版　　次:2016年1月第1版　印　次:2016年1月第1次印刷
印　　数:1~3000
定　　价:60.00元

产品编号:057309-01

# 目录

| | | |
|---|---|---|
| 自序："国际建筑工坊"十四年 | 程晓青　邹　欢 | 3 |
| "Atelier Internationale d'Architecture Construite" (AIAC) for 14 Years | Cheng Xiaoqing　Zou Huan | 9 |
| 国际学生建筑设计交流竞赛活动有感 | 栗德祥 | 16 |

| | | |
|---|---|---|
| 国际建筑工坊十四年教学活动分布 | | 002 |
| 哥伦比亚波哥大/安第斯大学国际学生中心 | | 004 |
| 　五彩安第斯 | 闵嘉剑 | 006 |
| 　双面波哥大 | 熊哲昆 | 012 |
| 日本东京/日本桥地区城市更新计划 | | 018 |
| 　河灯 | 张冰洁 | 020 |
| 　花忆 | 连晓刚 | 028 |
| 意大利威尼斯/大运河滨水建筑更新设计 | | 036 |
| 　城市百宝箱 | 陶一兰 | 038 |
| 　威尼斯花园 | 钟　琳 | 046 |
| 　缝合威尼斯 | 刘　芸 | 052 |
| 　环形客厅 | 孟　宁 | 058 |
| 法国奥什/军事遗迹改造与再利用 | | 060 |
| 　奥什复兴计划 | 朱　源　等 | 062 |
| 法国巴约纳/滨水空间更新计划 | | 074 |
| 　城市取景柜 | 李　昆　齐　际 | 076 |
| 　活力极＆舒适岛 | 刘伯宇　秦　岭 | 080 |
| 韩国晋州/泗川滨海现代音乐厅 | | 084 |
| 　梯田 | 戴天行　闫晋波 | 086 |
| 中国深圳/蛇口区滨海创意中心 | | 092 |
| 　垂挂城市 | 熊　星　秦嘉煜 | 094 |
| 　风之翼 | 李　烨　陈　忱 | 102 |
| 马来西亚马六甲/红树林湿地生态旅游中心 | | 110 |
| 　绿色半岛：城市绿点 | 张　烨　李　煦 | 112 |
| 　林中生活 | 国　萃　黄逾轩 | 122 |
| 法国马赛/军事遗迹改造与再利用 | | 130 |
| 　音乐节点连廊 | 兰　俊 | 132 |
| 　呼吸自然 | 李德熙 | 138 |
| 　活力社区 | 王　静 | 142 |
| 意大利巴勒莫/旧港区更新计划 | | 146 |
| 　滨海广场 | 霍振舟 | 148 |
| 韩国首尔现代艺术博物馆 | | 152 |
| 　走向文化与自然 | 雷　亮　申吉秀　郑　天 | 154 |
| 中国北京/ 高层生态写字楼 | | 158 |
| 　E空间 | 庞　聪　储以平　李　楠 | 160 |
| 法国索姆/滨海艺术家村 | | 166 |
| 　墙 | 陈　帆　等 | 168 |
| 附录1　国际建筑工坊历年教学统计 | | 174 |
| 附录2　国际建筑工坊参与院校统计 | | 176 |
| 致谢 | | 177 |

# 自序:"国际建筑工坊"十四年

## 1. 缘起:来自巴黎的美好邀约

"国际建筑工坊"的前身是由法国巴黎拉维莱特国立高等建筑学院 ERIC DUBOSC(埃里克·杜鲍斯克)教授与意大利米兰理工大学建筑学院 ETTORE ZAMBELLI(埃托·赞贝里)教授共同组织的两校学生建筑设计竞赛。2002 年,受 DUBOSC 教授的盛情邀请,清华大学建筑学院首次派出由栗德祥教授率领的师生团队参加该竞赛活动,同年参加的院校还有韩国汉阳大学建筑学院和意大利那不勒斯大学建筑学院。正是在 2002 年的活动中,各院校共同协商达成建立长期合作联盟的意向,并通过了合作框架,自此,"国际建筑工坊"的早期格局初步形成,由于当时成员共有四个国家的五所建筑院校,故被简称为"四国五校"联合设计。

"国际建筑工坊"自成立之初便确定了清晰的工作模式,即:以促进东西方建筑教育交流为主旨思想,参与的学生主要为各校一年级研究生,每年春季学期举办一次活动,由各院校轮流主持。合作和交流按照设计进程共分为两个阶段,设计前期各校师生共同前往负责出题的国家进行选址考察和学术交流,设计终期则共同前往负责展评的国家进行最终成果展览和优秀作品评选,选址考察和作品展评分别在两个国家进行,旨在为各国学生创造亲身感受不同建筑文化环境的机会。

"国际建筑工坊"的早期活动主要在中、法、意、韩的"四国五校"之间进行,随着时间的推移,这项活动以其独特魅力逐渐吸引了越来越多的各国院校加盟,德国、马来西亚、日本、哥伦比亚、秘鲁、越南、泰国和西班牙等国的优秀建筑院校纷纷加入为其带来了丰富的异域建筑文化思考,也为师生创造了更为广阔的交流空间。

十四年前,一个来自巴黎的美好邀约使清华大学建筑学院与"国际建筑工坊"结下不解之缘,究其原因是其主旨思想与清华积极拓展国际教育合作的理念不谋而合。作为创始成员之一,清华师生团队迄今已经连续十四次参加此项活动,"国际建筑工坊"从早期的课外选修设计竞赛,逐渐正式纳入清华建筑学专业研究生培养体系,到 2006 年成为必修设计课程的专题之一。在此过程中,我们与来自世界各地的同行共同探索建设领域的前沿课题,分享创新设计理念与心得,获益匪浅。

## 2. 探索:建筑教育合作新模式

自改革开放以来,清华大学建筑学院不断拓展国际教育合作,随着对外交流的加强,建筑教育层面的交流模式也在不断丰富:从 20 世纪 80 年代起步的公派访问学者,到 20 世纪 90 年代开始的学生交换培养,早期交流的重点是借助国外的教育平台,以单向对外输送人才为主,建筑专业教育的特点并不明显，受众也十分有限。21世纪以来，日益扩大的国际联合设计为建筑教育合作开创了新的模式，国际联合设计契合建筑学专业特点，以增强理解和相互学习为主，交流的重点转变为搭建中外平等的教育平台，输送与引进并举，受众人数大大增加，实现了研究生全覆盖。近年来，清华大学建筑学院国际联合设计不断扩展，在研究生教育中发挥出越来越重要的作用，究其原因与以下几个基本理念密不可分：

### · 合作原则——建立平等舞台

在"国际建筑工坊"漫长的建设过程中，对于不同文化背景的国家和院校始终以相互尊重为前提，摒弃基于国家大小、经济强弱、文化优劣的某些狭隘偏见，在其成员之中既有名列世界前茅的发达国家、知名院校，又有独具地域特色的发展中国家、新兴院校，参与国家总数达到12个，参与院校总数达到21所，平等是维系彼此友谊和稳定合作的基础。走访各个国家的每所院校，我们看到了迥异的教学环境和不同的教学方法，也看到了相同的专业追求和各自突出的学科优势，差异本身就是促进相互学习、共同进步的动力之一。平等的交流舞台为我们开拓了欣赏不同建筑文化精髓的视角，通过彼此的交流互补促进了教育理念的进步。

### · 关键难点——磨合教学理念

"国际建筑工坊"横跨欧洲、亚洲和拉美洲，参与国家和院校众多，各自的教学体系、授课时间和教学要求均有较大差异，对形成可持续的合作关系造成相当大的障碍。早期活动主要以较为松散的竞赛为主，学生自愿参加，成果不计入课程，对于激发学生积极性和深化设计方案非常不利。对此，各成员均就"如何协调活动时间？"、"如何将其纳入正规课程？"、"如何在共同命题之下发挥各校所长？"等关键难点进行了积极推进，也得到了各院校的大力支持。随着不同教学理念的相互磨合，逐渐建立了稳定和谐的合作关系，目前各主要成员院校均已将此项活动作为研究生必修课，并针对各自不同的教学特点，在完成共同设计成果之外深化相应的论文研究和技术设计等个性化内容。

### · 教学重点——开拓专业视野

作为一门研究生设计课程，"国际建筑工坊"的教学重点不是具体设计手法的传授，而是帮助学生提高综合分析问题的能力、建立科学的设计方法论，其中开拓专业视野是此项活动的最大收获。"国际建筑工坊"的广泛交流平台为学生们创造了感悟世界不同文化的机会，在每年的活动中都为学生组织两次学术交流互访，帮助他们深入了解建设环境的历史脉络和未来发展。教学活动并不局限于校园课堂和指导教师，而是广泛邀请以政府管理和规划部门、建设开发单位、当地代表建筑师等行业专家为学生们介绍城市发展的相关需求和进展，提供了十分难得的学习机会。学生们可以直观地了解不同城市的发展问题，亲耳聆听国际建筑领域前沿理论，亲身探究东西方建筑思想的异同，现场考察优秀设计案例，对于其形成全面的建筑观具有积极意义。

· 教学方法——强调激发互动

结合清华建筑学专业研究生培养目标，"国际建筑工坊"注重设计方法的传授，采用激发互动式的教学方法，以研讨课为主要教学形式。一方面，鼓励学生积极开展自主研究，从每个项目所在国家的历史文化和社会发展入手，锻炼其发现问题的能力；另一方面，教师按照教学进程设置阶段性的成果要求，帮助学生循序渐进地不断深化研究和设计，形成逻辑清晰的设计思路，增强其解决问题的能力；同时，引入多学科交叉的教学内容，促进学生方案的深化，提高其专业设计的能力。

· 核心目标——回归本土文化

"和而不同"是中国人对不同文化的独特理解，强调认同外来文化而不抹杀自身特点，二者相辅相成，缺一不可。清华大学建筑学院在"国际建筑工坊"教学中始终以异质文化之间的相互借鉴作为提高教学水平、促进文化自信的基本手段，交流的核心目标是回归本土文化，这一追求亦得到了每个成员院校的认同。在十四年的教学实践中，既强调对项目所处环境和地域文化的尊重，又鼓励学生不拘泥于建设条件，大胆提出基于自身理解的设计方案。随着教学思想的日渐成熟，清华学生所展现的兼具浓厚东方式思维和前瞻设计理念的作品赢得了国际同行的广泛赞许。在每年的成果展评中，我们非常欣慰地看到各院校师生所奉献的独具特色、风格迥异的设计理念，相信这些基于不同文化背景的另类思考对于每个国家和城市的发展都具有积极的启发价值。

## 3. 感悟：异质文化碰撞与升华

在"国际建筑工坊"十四年的实践中令人感触最深的是异质文化的碰撞，由于各所院校来自不同的文化背景，拥有不同的教育理念，面对同样的研究课题时往往会采用差异性极大的思维方法和解决方案。因此，认真分析不同的教学理念、积极学习彼此的实践经验，对于提升各自的教学水平十分有益。

· 共同的关注视角

"国际建筑工坊"自创立以来，始终关注城市和建筑发展的前沿问题，强调命题的前瞻性。在历年的设计选题中，各主办院校积极寻找城市和建筑发展的关键难点作为研究和设计题目，虽然环境不同、内容各异，但是都呈现出共同的关注视角，即以生态技术、城市与建筑更新和地域文化为重点的当代建筑核心命题。

早期阶段设计选题的突出特色是要求采用绿色建筑和节能技术，在完成基本设计方案的同时还须进行深入的技术模拟和节点设计，针对不同的地段条件，绿色技术的选择具有很强的针对性。例如：在 2003 年北京高层综合体设计中，积极探索了冬冷夏热地区高层建筑的被动式节能技术，方案针对保温、通风和隔热等进行了多角度的探索和模拟，对于亚热带环境的绿色建筑设计实践具有很好的参考价值。在 2007 年马六甲湿地居住建筑和 2008 年深圳滨海公共建筑设计中，潮湿闷热的气候特点对于方案的形成带来很大影响，架空、遮阳、风巷、水墙和植被维护结构等创新性解决方案对于推动热带环境的建筑节能研究具有积极意义。

城市与建筑更新是"国际建筑工坊"关注的又一个重点，无论是位于首尔景福宫一隅的现代艺术中心设计，还是贝约纳河畔的拿破仑兵营再造；无论是威尼斯大运河上的旧建筑更新，还是东京日本桥高架路的再利用，每个选题无不铭刻着各个国家的发展脉络、展示出各个城市的历史变迁。关注历史、关注文化、关注城市和社会发展加深了我们对于个体建筑定位的思考，整合既有空间秩序、实现新旧建筑的融合、再造美好的城市环境成为设计的重要指导思想，为激发每座城市的活力进行不懈努力。

随着参与国家和院校的不断增加，异质文化的影响令"国际建筑工坊"的选题内容日益丰富，特别是在地域文化特色方面的探索不断加强。历史悠久的欧洲古典主义建筑、秩序清晰的韩国宫殿和园林、独具特色的马来屋住宅、创意无限的拉美城市涂鸦艺术为设计者带来了丰富的创作源泉和艺术灵感。以差异化的视角审视异质文化，跨越不同文化背景的地域建筑探索往往最能"去粗取精"，收获启发性的思想火花。

## · 迥异的解决方案

成果展评是"国际建筑工坊"始终坚持的重要交流环节，每年数十个方案供师生彼此观摩，历经严格的评审选出最优秀的设计，坚持这一过程的核心目标就是通过展示在共同命题之下不同院校所提出的风格迥异的解决方案，促进教学理念和方法的相互学习。每组设计方案往往清晰地体现了各院校突出的学术优势，折射出鲜明的教学思想。合作院校各自的教学侧重点十分明显，例如：意大利米兰理工大学注重技术和建构层面的训练，方案的技术深度优势明显；威尼斯建筑学院则注重设计理念、特别是哲学层面的思考，方案的理念和表达手法纯熟；法国拉维莱特国立高等建筑学院在新型轻钢结构体系方面经验丰富，方案充满轻盈浪漫的气息；与之相比，亚洲各院校则更加关注历史脉络的梳理和城市尺度的思考，方案在突出地域性和现代感的矛盾中脱颖而出。

## · 相互碰撞中升华

在"国际建筑工坊"的同行评价中，清华团队在这些年中的进步是有目共睹的，从早期努力摸索，到慢慢适应规则，进而完美融入并发挥重要作用，经历了从被动到主动的角色转换，逐渐形成明显的学术优势。对此，作为教师我们深感自豪的同时，也清醒地认识到这是与整个国家的强盛和对外影响的扩展一脉相承的。早期的"国际建筑工坊"是清华学生了解国际建筑教育仅有的几个窗口之一，如今，随着学生国际视野的拓展、语言交流能力的提升、设计构思创造性的加强，"国际建筑工坊"逐渐成为展现清华学人风采的舞台。清华团队连续多年在成果评选中取得佳绩，这进一步提升了我们的文化自信。相信对于每一位参加此项活动的学生来说，"国际建筑工坊"所带给他们的启发和收获一定是受益终身的。

同样，在与国际同行的交流中，我们不断完善清华建筑学专业研究生教学体系，强调方法训练、开展专题研讨、跨学科合作和深化技术设计等教学改革一定程度上也是受益于"国际建筑工坊"的相互学习。

## 4. 收获：行走中的激励与成长

十四年来，我们仿佛一直行走在追求建筑理想的道路上，走过很多美丽的城市、看过很多奇异的风景，为大自然的鬼斧神工惊叹过、为人类创造的巧夺天工感动过，曾经的美好回忆今天仍旧历历在目：难忘西西里静谧的红砂岩建筑遗迹，难忘马来西亚柔佛州贫穷的水上人家；难忘文艺复兴之城的和谐优雅，难忘威尼斯"彩色岛"的色彩斑斓；难忘庆尚南道古寺的晨钟暮鼓，难忘福建土楼"婚礼"的热闹喧嚣；难忘壮美的马丘比丘，难忘宏伟的拿破仑兵营；难忘安第斯大学充满欢声笑语的校园生活，难忘 UNSCO 大厦神圣庄严的颁奖仪式……"国际建筑工坊"带给我们的收获是弥足珍贵的：

其一，作为教育者——通过"国际建筑工坊"的平台，我们广交各国朋友，收获了崭新的教学思想，提升了自身的教学水平。在历年的教学实践中，我们为这门课程倾注了太多的心血，每个设计的教学过程同样是学习过程，我们和学生们一起研究和分享，为每一位学生的困惑而担忧，为每一个方案的进展所鼓舞，教学理念在潜移默化中得以提升。

其二，作为建筑师——在"国际建筑工坊"中我们拓展了专业视野，交流前沿理论，分享创作心得。建筑学专业教育是一项特殊的事业，每位教师同时也是建筑师，在个人专业发展道路上，设计实践与理论教学二者缺一不可，所谓教学相长，教学理念方面的提升同样折射在我们作为建筑师的设计创作中。

其三，作为清华人——我们为学生们的出色表现而欢呼、自豪，是学生们的聪颖和智慧为人类的建设事业奉献了美好的创意与思想，是学生们的勤奋和努力为清华大学赢得了崇高的国际声誉。

最后，作为亲历者——我们见证了"国际建筑工坊"从艰苦创办到蓬勃发展的成长历程，来自世界各国的同行们心怀共同的专业追求，为了构建完美的教育交流平台不断摸索前行，在这段难忘的旅程中彼此激励与成长。

## 5. 愿景：开拓建筑教育的"丝绸之路"

时光荏苒，日月穿梭，不经意之间，"国际建筑工坊"迄今已经开展十四年了。十四年，对于人生来讲已是一段相当漫长的经历。同样，对于"国际建筑工坊"来说，随着成员不断增加、交流不断深入，仿佛一个咿呀学语的婴儿逐渐成长为意气风发的少年。回首这段难忘的历程，不得不承认我们的心中是怀有些许苦涩的：早期创始人之一的 ZAMBELLI 教授已经故去，为我们留下深深的怀念；曾经主持此项活动的部分中外教师也已退休，离开教学一线。然而，我们同时也看到越来越多的建筑学人加入到这一队伍中，他们带来了新的思想与活力。更加值得欣慰的是，曾经参与此项活动的学生们如今已经活跃在世界各地的建设领域，在各自岗位上积极实践着、创造着。其中甚至有的学生也已经步入教职，担当着推动"国际建筑工坊"扩大传承的重任。2015 年，"国际建筑工坊"首次在越南河内举行，参与的国家和学校进一步增多，学生总人数达到 200 多人，国际影响日益彰显。

"国际建筑工坊"的初衷是搭建东西方建筑教育合作的桥梁，努力开拓一条建筑教育的"丝绸之路"。

感谢十四年来陪伴我们一起进步的各国同行,这是一群对建筑充满热爱的人,他们为着无疆界的建筑理想,为着人类的共同发展而努力。应该说,通过此项活动令我们更加热爱这个美好的世界,也更加热爱所从事的伟大事业。

**祝愿我们的友谊长存!祝福"国际建筑工坊"薪火相传、蓬勃发展!**

程晓青　邹欢

2015 年 7 月于清华园

# "Atelier Internationale d'Architecture Construite" (AIAC) for 14 Years

## 1. Origin: A beautiful invitation from Paris

The predecessor of "AIAC" is an international architectural design competition jointly organized by Professor ERIC DUBOSC of l'Ecole Nationale superieure d'Architecture de Paris la villette and Professor ETTORE ZAMBELLI of Politecnico di Milano. In 2002, with the gracious invitation of Professor DUBOSC, the School of Architecture, Tsinghua University dispatched a team of teachers and students led by Professor Li Dexiang to participate in the activity. In the same year, the School of Architecture of Hanyang University, R.O.Korea and School of Architecture of Naples from Italy also attended the event. Through the activity in 2002, all participating schools agreed with an intention of establishing a long-term cooperative alliance through joint consultation, and a framework agreement was successfully passed. Henceforth, the early structure of "AIAC" came into being. As the members then were from five schools of four countries, the activity is therefore abbreviated as the Four Coutries Five Schools joint studio.

Early after its establishment, "AIAC" has confirmed a clear working method: to follow the key thought of promoting the exchange between the occidental and the oriental architectural education. Participants are mainly first-year graduate students. The activity is held once in each spring term, hosted by all schools in turn. According to the design progress, the cooperation and exchange take in two steps. In the first step, teachers and students from all schools will visit the selected country for site selection and academic exchange. In the second step, they will go to the country that is responsible for exhibition and appraisal to exhibit their final achievements and contest for excellent works. The investigation of site selection and work appraisal will be conducted respectively in two countries, so as to create an opportunity by which students of all countries can personally experience a different architectural and cultural environment.

The preliminary activities of "AIAC" were mainly conducted among the Four Coutries Five Schools, namely, China, France, Italy and R.O.Korea. As time goes on, this activity has aroused the interest of more and more Universities from other countries gradually, for its unique charm. The excellent architecture schools from Germany, Malaysia, Japan, Colombia, Peru, Vietnam, Thailand and Spain have also introduced rich ideas about foreign architectural cultures, which also create a broader space of exchange for teachers and students.

Fourteen years ago, the School of Architecture, Tsinghua University has been fated to have a permanent relation with "AIAC" since a beautiful invitation from Paris. One of the reasons is that the key thought of the activity coincides with Tsinghua's concept of actively expanding the international cooperation in education. As one of the founders, the team of teachers and students from Tsinghua has so far participated in fourteen sessions consecutively. "AIAC" has been gradually included in the graduate student pedagogic programme of Tsinghua's specialty of Architecture, developed from an extracurricular selective design competition at the early days. Till 2006, this has been one of the special topics of required design curriculum. In that process, we jointly explored the leading topics in the field of architecture and shared our creative design concepts and achievements with peers from all over the world, from which we have benefited a great deal.

## 2. Exploration: A new model of cooperation in architectural education

Since the reform and opening up, the School of Architecture, Tsinghua University has been expanding its cooperation in international education continuously. As the foreign exchange is strengthened, the exchange patterns at the level of architectural education have been enriched ceaselessly: From government-sponsored visiting scholars in the 1980s to the cultivation of student exchange in the 1990s, the early exchange focused on sending talents abroad unilaterally via the foreign educational platforms, so that the features of professional architectural education had not been highlighted yet and participants were quite limited. Since 2000, the increasing international joint design has ushered in a new pattern of cooperation in the architectural education. The international joint design agrees with the features of architecture, with an aim to enhance understanding and mutual learning. The emphasis of exchange is changed into the establishment of an educational platform that is fair for both China and foreign countries. The activity pays equal attention to both sending and introducing of talents and its participants are increasing sharply, realizing the full coverage for graduate students. In recent years, the School of Architecture, Tsinghua University has continuously developed the international joint design and it played an increasingly significant role in postgraduate education. These are inseparable of the following basic concepts:

- **The Principle of Cooperation - To establish an common stage**

During the long process of construction of "AIAC", countries and schools with different cultural backgrounds always abide by the premise of respecting each other and abandon certain prejudices on the size of country, economic strength and cultural context. Members include not only the world's developed countries and well-known schools but also developing countries and emerging schools with unique regional characteristics. A total of 21 schools from 12 countries have participated in this activity. Equality is the basis to sustain the mutual friendship and stable cooperation among members. In our visits to each school of each country, we saw different teaching environment and unique teaching methodology, and also the same pursuit for specialty and prominent disciplinary advantages. The difference itself is one of the driving forces to promote mutual learning and combined development. The common exchange stage has created a perspective for us to appreciate the essence of different architectural cultures. The complementarity in exchange gives an impetus to the advancement of educational concept.

- **The Key Difficulties - To coordinate teaching concepts**

"AIAC" covers a number of countries and schools in Europe, Asia and Latin America. The big difference in the system, time and requirements of teaching has imposed a considerably great barrier to the formation of sustained cooperative relationship. The early activities focused on the unconsolidated competition, students may participate in of their own will, but the achievements were not included in the curriculum. Consequently, this was quite unfavorable to stimulate the students' enthusiasm and to deepen their design scheme. On this ground, all members made dedicated efforts to tackle the key difficulties such as "How to coordinate the activity time?" "How to include it in the formal curriculum?" "How to exert the advantages of each school under the common proposition?" This also obtained the great support of all schools. A stable and harmonious cooperative relationship has been established gradually along with the mutual coordination among the different concepts. Currently, all key members have included this activity into the pedagogic programme of postgraduates, and deepened the corresponding thesis research, technical design and other personalized contents in addition to completing the joint design achievement, in view of different teaching features of each member.

- **The Teaching Priority - To develop a professional view**

As a design project of postgraduates, the teaching priority of "AIAC" is by no means to impart the specific design techniques, but to help students enhance their ability of comprehensively analyzing and establish a logical design methodology. In particular, the maximum benefit is to widen their professional view. The wide-ranging exchange platform of "AIAC" provides students with an opportunity to experience different cultures in the world. In the annual activity, we organize twice academic exchanges and visits for students, so as to help them profoundly understand the historical context and future development of construction environment. Teaching activities are not limited to lectures in campus and guidance from teachers, but to widely invite experts from the governmental administration and planning departments, construction and development units, and regional representative architects to introduce the local needs and progress of urban space. This has created a quite rare opportunity of learning. Students can have an intuitive understanding to the development issues of different cities, know the cutting-edge theory of the international architectural fields, explore the similarities and differences between Chinese and western architectural thinking, and make field inspection on excellent design cases. These aspects are of positive significance to form all-round architectural values.

- **The Teaching Methodology - To emphasize the stimulated interaction**

By combining the pedagogic objective of Tsinghua's architecture postgraduates, "AIAC" attaches importance to imparting the methodology and adopts the teaching method of stimulated interaction with seminars as the main teaching form. On the one hand, students are encouraged to actively make independent research and exercise their ability of problem finding by starting with history, cultures and social development studies of local country of each project; on the other hand, teachers impose the requirements for phased results according to the teaching schedule, so as to help students continuously deepen their research and design step by step, form the design concept with clear logic and boost their ability of problem finding; meanwhile, multi-disciplinary teaching contents are introduced to deepen the students' projects and enhance their professional design ability.

- **The Core Objective - To return to native cultures**

The "Harmony in diversity" represents the unique concept of Chinese people regarding different cultures; such concept suggests that people agree with foreign cultures but not obliterate their own characteristics. These two complementary aspects are indispensable of each other. On this ground, in the teaching process of "AIAC", the School of Architecture,

Tsinghua University has been always adhering to the basic approach of enhancing the teaching level and promoting the confidence in cultures by mutual interaction of heterogeneous cultures. The core objective of exchange is to return to native cultures, which is also recognized by each member school. During fourteen years of teaching practices, we not only emphasize the respect to the local environment and regional cultures but also encourage students to boldly propose a design scheme based on their own understanding, instead of rigidly adhering to the construction conditions. As the teaching ideas are matured increasingly, Tsinghua's students have manifested the works that boast rich oriental ideas and prospective design concept, and thus won the widespread appraisal among the international peers. In the annual exhibition and appraisal of results, we are delighted with the design concepts of teachers and students of all schools. Their works are in possession of unique features and different ways. We believe that these different thinking based on different cultural background is of positive enlightenment for the development of each country and each city.

## 3. Perception: Collision and sublimation of heterogeneous cultures

In fourteen years of practices of "AIAC", what we have most realised is the collision of heterogeneous cultures. As those schools have different cultural context and teaching concept, we generally adopt the thinking methods and solutions with great difference in face of the same research subjects. Hence, to earnestly analyze different teaching concept and actively learn the practical experience from each other is of great benefit to enhancing each other's teaching achievement.

• **Viewpoints in Common**

Since its establishment, "AIAC" has always paid attention to the frontier questions about the development of cities and architecture, and emphasized the perceptiveness of propositions. In the previous selected subjects, all sponsoring schools actively sought for the key difficulties facing the development of cities and architecture as the assignments of research and design. Despite different environment and contents, all those subjects shared viewpoints in common, specifically, taking ecological technologies, urban renewals and regional cultures as the core objectives of contemporary architecture.

At the early stages, the design subjects were required to highlight green building and energy-saving technology. While completing the basic design scheme, the selected subjects must also go further into technical simulation and detail design. Upon the selection, the green technology shall be especially specific to the different conditions of a site. For example, Beijing high-rise complex design in 2003 manifested an active exploration into passive energy conservation technology for high-rise buildings in areas which are hot in summer and cold in winter. The scheme made a multi-dimensional exploration and simulation specific to heat preservation, ventilation and heat isolation. This had good reference value for practices of green building design in the subtropical environment. In the design of residential buildings for Malacca Wetland in 2007 and public building design of Shenzhen in 2008, the warm and wet climate features had a big impact on the projects. Hence, the innovative solutions such as building on stilts, sunshade, ventilation alley, waterwall and vegetation maintenance structure played an important role in promoting the research into building energy efficiency in the tropical environment.

Another key concern of "AIAC" is the renewal of cities and buildings. Whether the design of modern arts center in the neighbour of Gyeongbokgung Palace in Seoul or reconstruction of Napoleon barracks in the riverside of Bayonne, and whether the renewal of old buildings along the Grand Canal in Venice or reutilization of elevated highway in Nihonbashi of Tokyo, each of those topics, without exception, took on the development context of the whole country and demonstrated the historical changes of each city. The concerns to history, culture, city and social development have deepened our thinking about individual architectural performance. One of the important guiding ideologies for teaching is to integrate the existing spatial patterns, realize the combination of new and old buildings and reconstruct the beautiful urban environment. Following such an ideology, we will make unremitting efforts to activate the vitality of each country.

As the participating members are increasing, the subjects of "AIAC" affected by heterogeneous cultures are becoming richer and richer. In particular, the exploration in the regional cultural features are strengthened continuously. The time-honored European classicism buildings, orderly palaces and gardens of R.O.Korea, characteristic Malaysian residential buildings and creative Graffiti arts in Latin American cities, have brought rich source of creation and artistic inspiration. It is most likely to "eliminate all but the essence" and harvest the enlightening sparks of thought by carefully examining the heterogeneous cultures from a differentiated perspective and explore the regional buildings under different cultural contexts.

- **Disparate solutions**

The exhibition and appraisal of designing results constitute an important link of exchange that "AIAC" has always adhered to. Every year, dozens of projects are assembled for inspection and learning from each other. Through the strict appraisal and review, the most excellent designs will come out. However, the objective of this process is to promote participants to learn from each other's teaching concepts and methodologies by the exhibition of those unique solutions as proposed by different schools under the same subject. Each group of project often clearly embodies the prominent academic advantages of each school and reflects their bright ideas. All schools in the cooperation have a quite obvious emphasis of pedagogy. For instance, Polytechnic University of Milan in Italy pays more attention on the trainings of techniques and construction levels, whose projects manifest obvious technical advantages; University IUAV of Venice focuses on thinking about design concept, especially the philosophical concept, and the project boasts sophisticated concept and means of expression; ENSAPLV boasts rich experience in light steel structure system, whose project is full of lightness and romantic atmosphere; by contrast, all schools in Asia focus far more on reflection the historical context and thinking about dimension of the city, and the final results stand out of the contradictions between regional characters and sense of modernity.

- **Sublimating in the collision**

In the peer evaluation of "AIAC", the progress of Tsinghua's team in these years is well known. From struggling exploration at the beginning to adapting itself to the rules slowly, Tsinghua's team has integrated itself into the activity perfectly and played a significant role. Through the transition from a passive to an active role, this team is gradually forming its obvious academic advantages. While we as teachers feel proud of it, we also realize clearly that this is closely related to the prosperity of our nation and its influence abroad. In the early days, "AIAC" was one of the few windows through which Tsinghua's students could know the international architectural education. However, nowadays, "AIAC" has gradually become a stage of reflecting the elegant demeanor of Tsinghua's scholars along with the widened international view of students, the improving ability of language communication and the improved design concept creativity. Tsinghua's team has obtained good results in the appraisal for years consecutively, which has further enhanced our cultural confidence. We believe that, for each student participating in this activity, the enlightenment and achievements they have obtained from "AIAC" will benefit them for life.

Similarly, we continuously improve the postgraduate teaching system of architecture of Tsinghua University during the exchange with international peers. To a certain extent, our reforms in emphasizing the methodological training, conducting seminars and interdisciplinary cooperation, and deepening technical design also benefit from mutual learning via "AIAC".

## 4. Achievements: Incentives and growth on the way

In fourteen years, it appears that we have always walked in the path of pursuing our architectural ideal. We have been to many beautiful cities, seen numerous fantastic sights, shocked by superlative craftsmanship of the nature and moved by wonderful workmanship of human beings. The beautiful memory is unforgettable and still visible: the tranquil red sandstone building ruins in Sicily and the poor boat dwellers in Johor, Malaysia; the harmony and elegancy of the Renaissance city and colorful "Burano Island" of Venice; morning bell and evening drum in Gyeongsangnam-do old temple and joyful "wedding ceremony" of Fujian Tulou; the most majestic Machu Picchu and the magnificent Napoleon barracks; campus life full of joy and laughter in University of the Andes, the sacred and solemn awarding ceremony in the UNESCO mansion... The achievements we've made through "AIAC" are pleasurable and valuable:

Firstly, as educators - we have made friends from all countries, learned fresh pedagogic ideas from them and enhanced our own via the platform of "AIAC". In all our teaching practices, we have made much more efforts in this course. Each teaching process we design is also a learning process. We experience and share it together with our students. We were concerned about each student's confusion and we were inspired by the progress of each project. Our teaching concept has been enhanced in the subtle influence.

Secondly, as architects - we have widen our professional view, exchanged the cutting-edge theory and shared our creation in the activity of "AIAC". The specialty education of architecture is a special undertaking. Each teacher is also an architect. On the way of personal development, design practices and theoretical teaching are indispensable of each other, as the saying goes that "teaching and learning benefit each other". The enhancement in pedagogic concept is similarly reflected in the design creation of architects.

Thirdly, as members of Tsinghua- we cheer for and feel proud of the outstanding performance of our students. They have contributed beautiful creativity and ideas for humans' architectural undertaking through their intelligence and wisdom. Their diligence and endeavors have won the great international fame for Tsinghua.

At last, as witnesses- we have witnessed the growth history of "AIAC" from tough establishment to flourishing development. With the common professional pursuit, the peers from all over the world have continuously explored how to build a perfect educational exchange platform. During this unforgettable journey, we have encouraged each other and grow up.

## 5. Vision: To construct the "Silk Road" of architectural education

Time flies like an arrow. So far, "AIAC" has been held for 14 years. Fourteen years, for life, represents a considerably long experience. Similarly, for "AIAC", it seems that a babbling infant is growing with lively spirits as the members have been increasing and exchange deepened. Looking back at this unforgettable process, we haveto acknowledge that we indeed felt bitter and astringent somewhat: Professor ZAMBELLI, one of the early founders, passed away, leaving us in deep yearnings; part of domestic and foreign teachers who hosted this activity have retired and left from classroom. However, in the meantime, we have seen that more and more architectural scholars are joining us, and they bring new thought and vigor. It is encouraging to know that students who once participated in this activity have been active on the architectural field in the world, and they are actively practicing and creating on their respective post. Even some of the students have returned to their homeland to serve as teachers and taken the heavy responsibility for promoting and inheriting "AIAC". In 2015, "AIAC" is held in Hanoi, Vietnam for the first time, with the increase of participating countries and schools. With More than 200 students participating in the activity, its international influence has been more and more apparent.

The original intention of "AIAC" is to establish a bridge occidental and oriental cooperation in the architectural education and aim to develop the "Silk Road" for architectural education. We would like to appreciate it that peers of all countries have witnessed our common advancement. This is a group of people who have love for architecture deeply. With the borderless architectural ideal, we are making efforts to achieve joint development of human beings. We might say that, through this activity, we love this beautiful world more and more, and it is also the great cause we are striving for.

May our friendship last forever! And "AIAC" is sincerely wished to grow vigorously generation by generation!

Cheng Xiaoqing   Zou Huan<br>
Tsinghua University, July 2015

# 国际学生建筑设计交流竞赛活动有感

"国际建筑工坊"活动起始于2002年春,主要内容是国际学生建筑设计竞赛,每年一届,每届春秋两次集中交流,分别在欧洲、亚洲或其他洲。春季交流一般在三月份,主要内容是发布和解释设计竞赛题目,调研建设地段设计条件和环境特点,听取指导老师们的学术报告,并参观考察地段所在地区的城市和建筑。秋季交流一般在九月份,主要内容是各国学生观摩设计成果,由指导老师和当地著名建筑师组成评委会评出获奖作品并颁奖,顺便参观考察该地区的城市和建筑。

"国际建筑工坊"活动从2002年起迄今已成功举办了14届,参加的国家和院校由最初的"四国五校"(法国拉维莱特国立高等建筑学院、意大利米兰理工大学、中国清华大学建筑学院、韩国汉阳大学建筑学院和意大利那不勒斯大学建筑学院)逐年有所拓展,迄今为止参与国家共12个,参与院校达21所。早期活动主要在法国、中国、意大利、韩国几个创始国举行,近年来逐渐扩展到在马来西亚、日本、哥伦比亚和越南等国举行。

参与"国际建筑工坊"活动有两点体会比较深刻:

一、据我理解,与以往的国际竞赛不同,该活动强调交流和学习,并不刻意追求拿大奖,而是锻炼同学们巧妙解决问题的能力,设计出有特色的好建筑。无论是设计过程中还是评奖颁奖时刻,大多数师生都能保持平和的学习心态,这是最有利于学习的氛围。本着这样的心态我院师生收获良多。

其实,各校交流竞赛团队各有所长,值得汲取。比如,米兰理工大学团队参赛学生是建筑技术专业,建筑技术方面知识丰富,表达娴熟,建筑构思活跃。法国拉维莱特国立高等建筑学院团队强调建筑形态的创新和轻钢结构的应用,具有思维理性与表达浪漫的特征。韩国汉阳大学团队的方案多数布局紧凑,空间丰富多变,造型简约典雅,模型精致爽眼。清华大学建筑学院团队则注重对地段环境的调查研究,突出建筑与环境的整体性以及建筑形态的时代感与地方性。

二、有机会体验国际经典城市和建筑,对于在校学生来说是弥足珍贵和必要的,这项活动就提供这一契机。这些年来我们带领学生考察了意大利、法国、韩国、西班牙和马来西亚等国的众多城市和建筑。

体验城市和建筑重点在城市总体风貌、街道尺度、建筑特色和景观绿化。所以每到一个城市，必先找到城市制高点，如大教堂的采光亭、钟楼或高地，老师带头爬上去，学生自然跟上来，俯瞰环视，城市总体概貌了然于胸，回到地面再细看街道、建筑和景观。我们参观体验过的欧洲城市都有某种魅力和活力。魅力主要体现在：城市与自然生态环境友好相处，空间布局既有多样性又有秩序感，公共空间及建筑有亲切的尺度感，有活跃的街道生活。历史遗存受到全面保护，地方文化得到弘扬，市民具有良好素质。其活力主要体现在商业、文化、休闲娱乐等消费领域的繁荣。意大利的小城锡耶纳就是这方面的典型，学生参观意大利城市多半少不了锡耶纳，它也特别受到中国建筑师们的青睐。

我们希望，通过参与"国际建筑工坊"活动，为同学们的人生旅程增添美好的回忆，为今后的职业生涯留下有益的参照。

栗德祥

2015 年 7 月

国际建筑工坊 2002—2014 年

# 国际建筑工坊十四年教学活动分布
# PEDAGOGIC ACTIVITIES' LOCATIONS OF AIAC IN THE PAST 14 YEARS

国际建筑工坊十四年教学活动分布 | 003

2014 年

# 哥伦比亚波哥大 / 安第斯大学国际学生中心
# INT. STUDENT CENTRE OF LOS ANDES UNIV. BOGOTÁ, COLOMBIA

指导教师 程晓青 邹欢 / 学生 闵嘉剑 熊哲昆 张华西 李金泰
Teacher: Cheng Xiaoqing, Zou Huan
Student: Min Jiajian, Xiong Zhekun, Zhang Huaxi, Li Jintai

2014年，"国际建筑工坊"第一次踏上南美大陆，来到了哥伦比亚首都波哥大。波哥大具有南美大都市的典型特征：紧密、混杂、动静对比鲜明、色彩鲜艳热烈，独特的地理特点体现在城市与自然的关系中：山脉成为城市的标志景观。本年度的题目是在安第斯大学校园附近设计一幢学生公寓。在对城市和地段进行了详细考察和调研后，学生们从城市街区生活复兴和活化的角度切入，力求植入标志性建筑以提升城市景观的价值。哥伦比亚建筑传统中独特的建筑材料——红砖为设计带来了灵感。用建筑解决城市社会问题也成为2014年"国际建筑工坊"广泛讨论的议题。

In 2014, AIAC sets foot on the South American Continent for the first time and arrived in Bogota, the capital of Colombia. Bogota boasts the classic features as a metropolis in South America: The city is compact, mixed and bright-colored with static and dynamic contrast. Unique geographical features are also reflected in the relationship between city and the nature: The mountain chain becomes a landmark landscape of the city. The project for students was to design a student apartment near the campus of University of the Andes. After carefully analysis and survey on the city and the selected site, students tried to integrate a landmark building into the city to enhance the value of civic landscape, from the perspective of reviving and enlivening the street life of the city. Red brick - a building material unique to the Colombian building traditions also inspired the design. A mostly discussion of 2014 AIAC was to make buildings as a solution to social problems of the city.

参与院校：
清华大学（中国）
巴黎拉维莱特国立高等建筑学院（法国）
威尼斯建筑大学（意大利）
汉阳大学（韩国）
庆尚大学（韩国）
庆应义塾大学（日本）
安第斯大学（哥伦比亚）
利马大学（秘鲁）

获一等奖院校：威尼斯建筑大学
清华获奖情况：闵嘉剑荣获优秀奖

# 五彩安第斯 / COLORFUL ANDES

指导教师 程晓青 邹欢 / 学生 闵嘉剑 / 2014 年
Teacher: Cheng Xiaoqing, Zou Huan / Student: Min Jiajian / 2014

波哥大位于安第斯山脚下,这座山脉是当地重要的景观、生活方式以及文化认同的象征,因此,创造出一种在山上居住的体验成为设计的重要的手段。原有的停车场被安排在地下,地面层变为开放的地景式的公园,与商店、咖啡厅等功能相结合。宿舍功能被架空在上方,形态呼应山的走势,建筑色彩呼应老城五彩的社区,最终形成一个富有活力的社区。

The Andes Mountain has always been an improtant symbol for Bogota, as a sight, a lifestyle and identity. Therefore, the project focuses a lot on the creation of an atmosphere of living in Andes. The original parking lot is arranged underground, with a new open park for the public and services. The dormitory is put above the ground, the shape of which corresponds with that of the mountain. Beautiful bright colors are used on the façade. Finally, it is expected to become a dynamic community.

# 五彩安第斯 / COLORFUL ANDES

### 安第斯山 MOUNTAIN ANDES

Mountain as SIGHT

Mountain as CULTURE

### 色彩分析 LA CANDELADIA COLOR RESEARCH

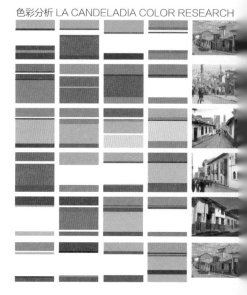

### 总图 SITE PLAN

### 住宅与社区空间构成 / HOUSING STRATEGIES

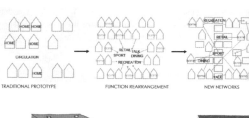

TRADITIONAL PROTOTYPE → FUNCTION REARRANGEMENT → NEW NETWORKS

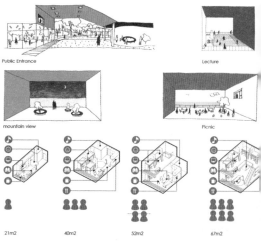

Public Entrance | Lecture
mountain view | Picnic

21m2 | 40m2 | 52m2 | 67m2

### 标准层平面 TYPICAL FLOOR PLAN

## 面向市民的公共空间
## PUBLIC SPACE FOR LOCAL RESIDENTS

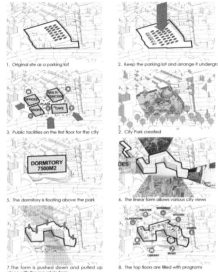

## 概念生成 CONCEPT GENERATION

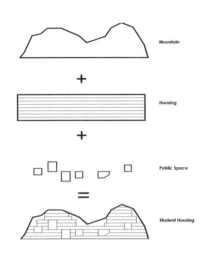

## 构成元素 STRUCTURE ELEMENTS

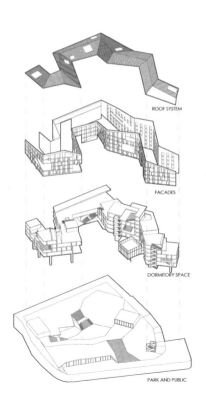

## 透视 PERSPECTIVES

# 五彩安第斯 / COLORFUL ANDES

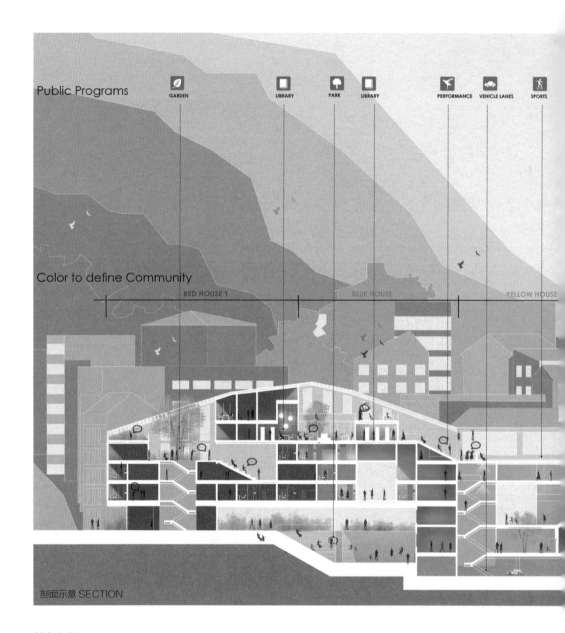

剖面示意 SECTION

**教师评语**

方案充分考虑了地段上重要的城市景观，以山和城市之间的连线作为贯穿建筑的视觉走廊，试图将城市公共空间与建筑内部交流空间联结起来，形成多层次的开放空间。连续的建筑体量功能合理，交通便捷。同时，通过色彩的运用呼应波哥大的城市特征。

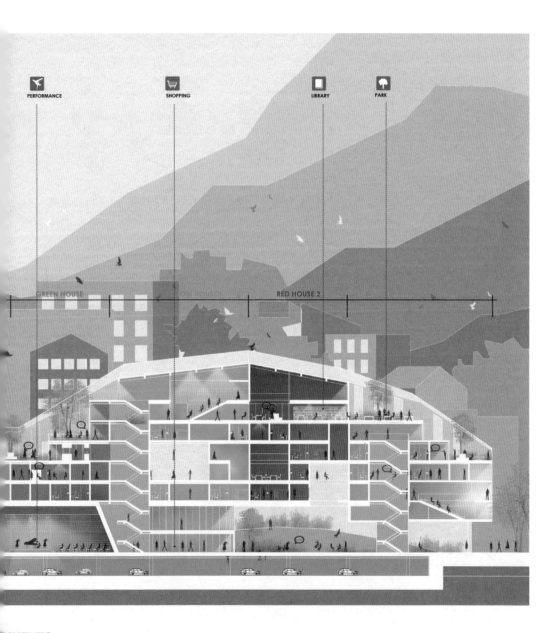

## COMMENTS

The project gives full consideration to the key visual corridors of urban landscape on the site and attempts to combine the public space and internal communication space of buildings to form a multi-level open space. The continuous architectural massing provides reasonable functions and convenient traffic. The application of colors responds to the urban features of Bogota.

# 双面波哥大 / BOGOTA IN DUALITY

指导教师 程晓青 邹欢 / 学生 熊哲昆 / 2014 年
Teacher: Cheng Xiaoqing, Zou Huan / Student: Xiong Zhekun / 2014

2014年/哥伦比亚波哥大/安第斯大学国际学生中心 | 013

波哥大：新与旧 BOGOTA: NEW & OLD

# 双面波哥大 / BOGOTA IN DUALITY

地处哥伦比亚首都波哥大的新旧城边界,这座学生公寓试图探索一种新的城市建筑形态——介于传统与未来、纪念与创新之间的"意外之城"——正如哥伦比亚伟大文学家马尔克斯闪烁着希望和忧郁的双重气质一样。在记忆迷失的新城,从未来回望过去;在绿色稀缺的老城,从过去远眺未来。一座活力四射的昨日故乡,一片城市共享的明日森林,出乎意料的校园生活、人际沟通、艺术创意在此激生蔓延。

如何在高密度的老城建设一座公园?它是一个空中的森林和散步系统,公园与半公共的庭院连接,形成充满活力的共享系统。如何在新城延续老城记忆?它是由空中广场、街区、标志物等组成的垂直城市意象,也是老城丰富活动新的载体。学生与市民在建筑中相遇,使它成为一座城市公园、空中运动场、涂鸦博物馆、室外音乐厅。老城中的旷野,科律中的叛逆。我们缅怀过去,面朝大海。

Situated in Bogota, Colombia, at the edge of the new town and the old town, the school student residence building is seeking an unexpected city form in between tradition and future, memory and innovation, just like the duality of the great Colombian novelist Gabriel José Márquez. A traditional hometown with great energy; a future park shared with the city. It provides green scenery for the dry old town, while reminding the new town of old treasures.

The project tries to ignite imagination through juxtaposition and collision of the duality of Bogota. It is a place where new form of meeting, communications and art innovation arises. We are seeking wildness from high-density, spontaneity within order. Saluting the past, and waving the future.

**教师评语**

方案构思大胆,尝试用巨构形成城市标志性建筑,并通过截然不同的立面处理隐喻地段所处环境的矛盾性,呼应城市街区的不同特征,为大学生创造乌托邦式的建筑体验。

**COMMENTS**

This is a bold concept which attempts to apply a mega structure to form the city's landmark building and hint the location of the site by different treatment of facades. It echoes with different features of urban blocks and creates a Utopia for students.

理念分析 CONCEPT ANALYSIS

城市 - 自然 CITY-NATURE: Populousness to go wildness

之间：介质 IN-BETWEEN: Intersticial Space

自然 - 城市 NATURE-CITY: Willdness to go Populousness

# 双面波哥大 / BOGOTA IN DUALITY

入口透视 ENTRANCE

技术设计 TECHNICAL DESIGN

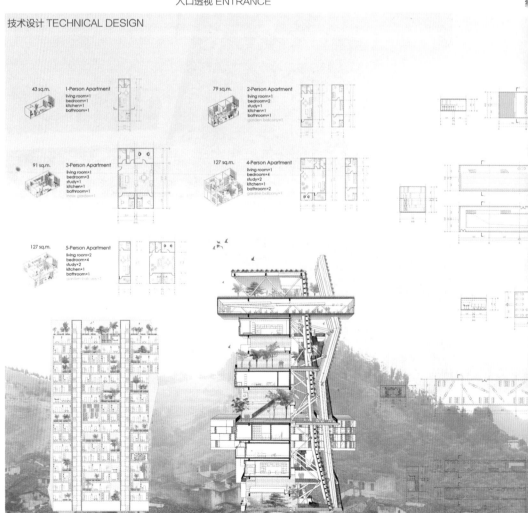

# 2014年/哥伦比亚波哥大/安第斯大学国际学生中心

公寓剖面 HOUSE SECTION

表皮设计 SLIDABLE LOUVRES

个性化采光 PERSONALIZED DAYLIGHTING

2013年
# 日本东京 / 日本桥地区城市更新计划
# URBAN RENEWAL OF NIHONBASHI REGION TOKYO, JAPAN

指导教师 程晓青 邹欢 / 学生 张冰洁 连晓刚 王若凡 邓施莹 李明扬 周南 茹笑岚 龚梦雅
Teacher: Cheng Xiaoqing, Zou Huan
Student: Zhang Bingjie, Lian Xiaogang, Wang Ruofan, Deng Shiying, Li Mingyang, Zhou Nan, Ru Xiaolan, Gong Mengya

作为2020夏季奥运会的主办城市，东京需要新的面貌。当年为承办1964年夏季奥运会，东京在城市中心修建了贯穿城市的高架路，解决了交通的问题，但是对城市景观和文物建筑造成了毁灭性的破坏。地段位于东京地理意义中心——日本桥的桥头，这里是东京的起点和起源。如同所有大都市在现代化过程中所经历的一样，技术决定论下的城市建设无法承载城市本质中所需要的文化生活，曾经引以为豪的高架路面临被拆除改造的命运，寸土寸金的中心地段需要引入更加高效、复合、充满活力的空间。围绕高架路拆除后滨河景观恢复和街道空间的再次繁荣，学生们对方案进行了多种方向的尝试，试图从历史文化、后信息社会及人本主义的角度，对这样一个错综复杂的城市地块的轮回再生给出答案。

As the host city of the 2020 Summer Olympic Games, Tokyo needs a new image. To hold the 1964 Summer Olympic Games, Tokyo built an elevated highway in the urban center of Tokyo, crossing the city. This road solved the traffic problems, but it brought about destructive effects on civic landscape and heritage buildings too. The site was located at the bridge head of Nihonbashi - the geographic center of Tokyo. This is the origin of Tokyo. As what a metropolis had experienced in the process of modernization, the city construction under the technological determinism couldn't undertake the cultural life required by the nature of city. The elevated highway that people took pride in faced the destiny of demolition, and the central site with scarce and expensive land needed a space that was more efficient, complex and full of activities. Around the riverside landscape restoration and re-prosperity of the streets after the demolition of the elevated highway, students made multiple attempts in their designs, and tried to give an answer to the reinstatement of such a complicated urban land parcel, from the perspective of history, culture, post-information society and humanism.

参与院校：
清华大学（中国）
巴黎拉维莱特国立高等建筑学院（法国）
威尼斯建筑大学（意大利）
汉阳大学（韩国）
庆尚大学（韩国）
庆应义塾大学（日本）
哈尔滨工业大学（中国）

获一等奖院校：巴黎拉维莱特国立高等建筑学院
清华获奖情况：张冰洁荣获二等奖
　　　　　　　连晓刚荣获优秀奖

# 河灯 / RIVER LANTERN

指导教师 程晓青 邹欢 / 学生 张冰洁 / 2013 年
Teacher: Cheng Xiaoqing, Zou Huan / Student: Zhang Bingjie / 2013

日本桥地区是江户时代繁华的商业中心,而今面临如下问题:高架桥覆盖河面,滨河空间消极;大型写字楼入侵,原住民外流;新综合中心兴盛,旧中心衰败;此区域密集紧凑,无空置土地。设计着力于激活滨河空间,使其成为休闲放松的特色后花园,重聚人气;然后微循环式插入办公、居住和商业设施,并在不同高度形成连续的散步道;最后在拆除高架后保留的基础上营造出活力而鲜美的河灯,代表城市的生活,寓意节日的庆典。

Nihonbashi was once the prosperous commercial center during the Edo period, while facing such problems nowadays: Highline makes along-river space negative. Giant office buildings force aborigines to leave. New centers leave the old center abandoned. The dense area provides few vacant lands.
The design tries to turn the along-river space into a featured public backyard to relax, regaining its popularity. Then office, residence and commercial facilities are carefully introduced, and promenades created on various levels. Finally lively and beautiful "river lanterns" are designed on former bases on which the demolished highline has stood. It is the urban life. It is the festival celebration.

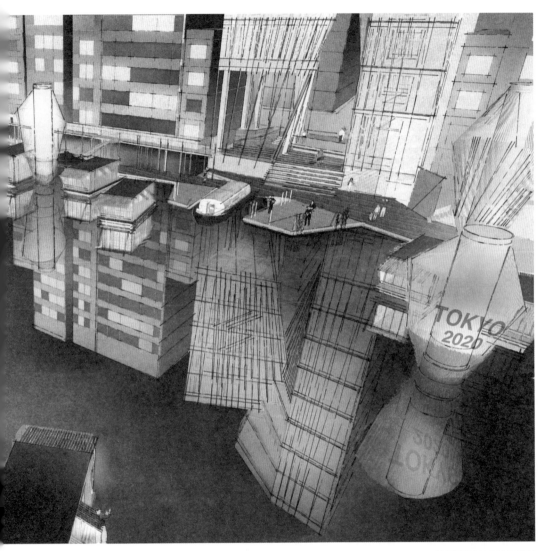

滨河透视 RIVER BANK

概念阐释 CONCEPT

# 河灯 / RIVER LANTERN

概念生成 CONCEPT GENERATION

环境分析 SITE ANALYSIS

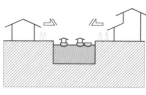

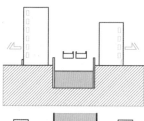

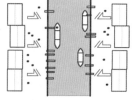

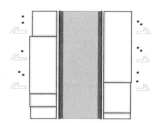

第一步：激活河岸空间
STEP 1
REVIVE THE RIVERBANKS

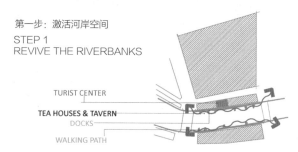

TURIST CENTER
TEA HOUSES & TAVERN
DOCKS
WALKING PATH

第二步：激活重点环境
STEP 2
REVIVE THE LEST ATTRACTIVE AREA

HOUSES
COMMERCIAL
WALKING PATH TO THE RIVERBANKS
CULTURAL EVENTS

第三步：激活新居住
STEP 3
MORE HOUSES, CONNCETED BY PATH

WALKING PATH TO THE RIVERBANKS

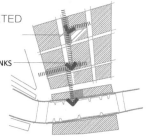

2013年/日本东京/日本桥地区城市更新计划 | 023

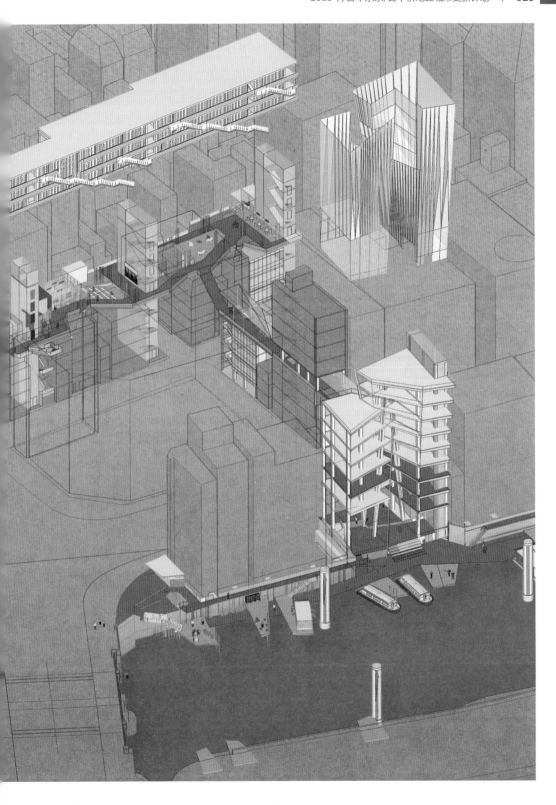

# 河灯 / RIVER LANTERN

F9

F8

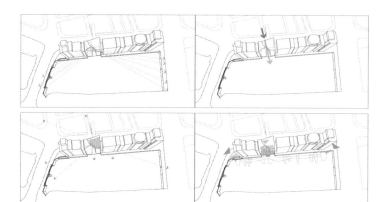

综合分析 ANALYSIS OF DESIGN

F7
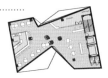

环境透视 PERSPECTIVE

F6

F5

F3

F2

办公部分各层平面 OFFICE PLANS

防坡堤改造 TETTY RENEWAL

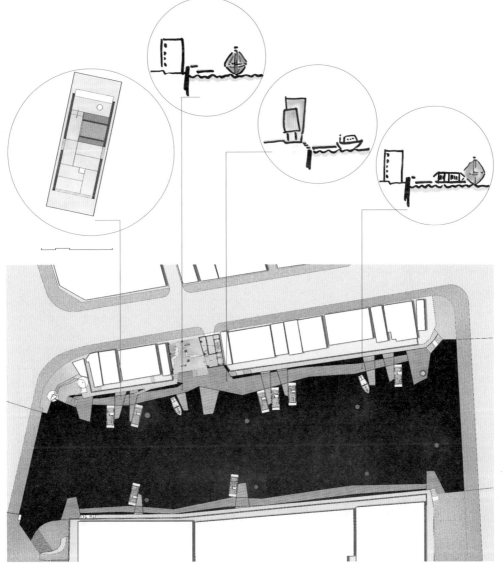

# 河灯 / RIVER LANTERN

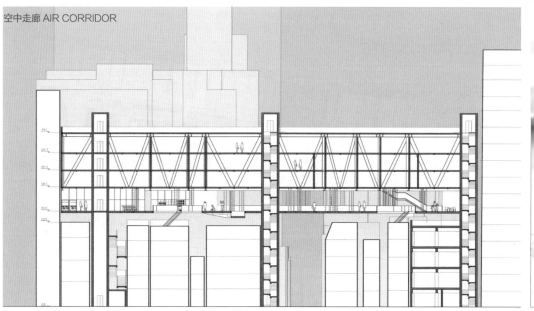

空中走廊 AIR CORRIDOR

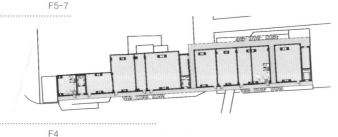

F5-7

F4

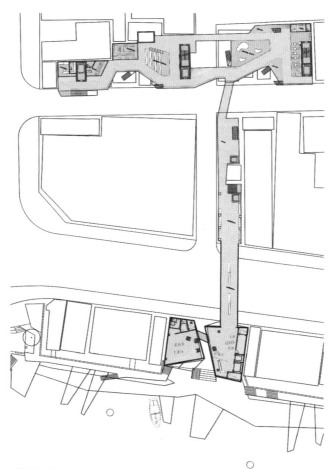

住宅平面
RESIDENCE PLANS

### 教师评语
传统节庆的城市景象启发了设计灵感，通过夜景与滨水建筑河面倒影来激活城市公共空间，和风浪漫，灵动巧妙。

### COMMENTS
The atmosphere in traditional festivals gives an idea. The urban public space is activated by night scenes and inverted image of waterfront buildings in the river, highlighting the romance, intelligence and ingenuity under the gentle breeze.

# 花忆 / URBAN GREENING

指导教师 程晓青 邹欢 / 学生 连晓刚 / 2013 年
Teacher: Cheng Xiaoqing, Zou Huan / Student: Lian Xiaogang / 2013

地段位于东京历史中心"日本桥"地区,计划在此对日本桥町做城市改造,并建造一座集商业、居住与办公功能的综合体。

The site is in the historical center "Nihon bashi" area in Tokyo. It is planned to have a holistic urban regeneration in this area. A complex with commercial, residential and living space will be built.

**教师评语**

建筑非常谨慎地与地段环境融为一体,原有高架桥改造为散步公园与建筑裙房连为一体,创造出丰富流动的城市滨水公共空间。

## COMMENTS

The building is cautiously integrated into the environment of the site. The Original viaduct is redesigned into a pedestrain garden, which, being an integral with the building complex, creates a colorful and flowing urban waterfront public space.

 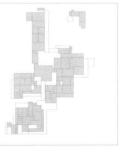 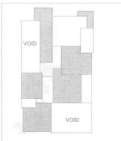 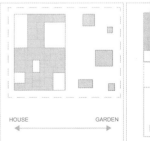

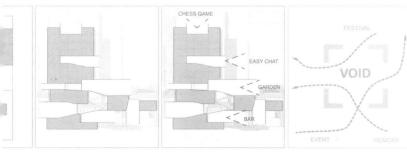

空间构成分析
SPACE ANALYSIS

# 花忆 / URBAN GREENING

城市变迁
EVOLUTION

645　　1603　　1868　　1911　　1945　　1964　　2013

城市绿化
URBAN GREEN

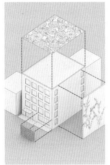

高架路再利用
HIGHWAY REUSE

屋顶绿化 ROOF GREENING

总图 SITE PLAN

公共空间　办公入口　停车入口
青旅入口
地铁入口

L GREENING　　　　　　　　　花架 PERGOLA

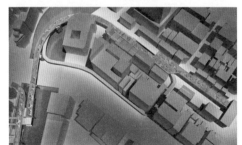

# 花忆 / URBAN GREENING

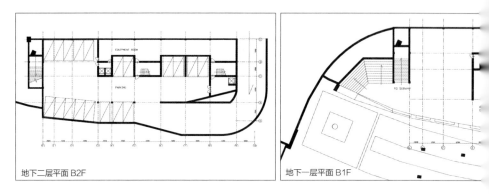

地下二层平面 B2F 　　　地下一层平面 B1F

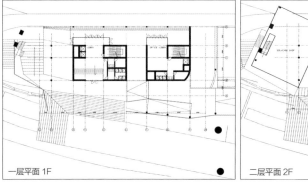

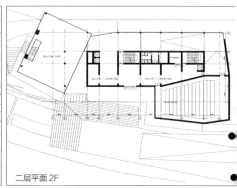

各层平面  
FLOOR PLANS

一层平面 1F 　　　二层平面 2F

功能分布  
FOUNCTION

1~3 层为公共空间，可自由抵达。4~10 层为青年旅社，各层空中花园通过室外楼梯连通。右图绿色部分示意高层的户外空间。
Floor 1~3 are public space. Floor 4~10 are hostel rooms with coutyards. These outdoor gardens are connected with stairs. The green areas in the following images are the gardens.

四层平面 4F 　　　五层平面 5F 　　　六层

花的节目  
FESTIVAL OF FLOWER

JULY. 31  
Nihonbashi Cleaning Day  
橋を洗う

AUG. 15  
Latern Festival  
お盆

JUL.15 – AUG.31  
2020 Tokyo Olympic  
2020 東京オリンピック

OCT. – NOV.  
Chrysanthemum  
菊まつり

NO°  
Ma  
秋祭

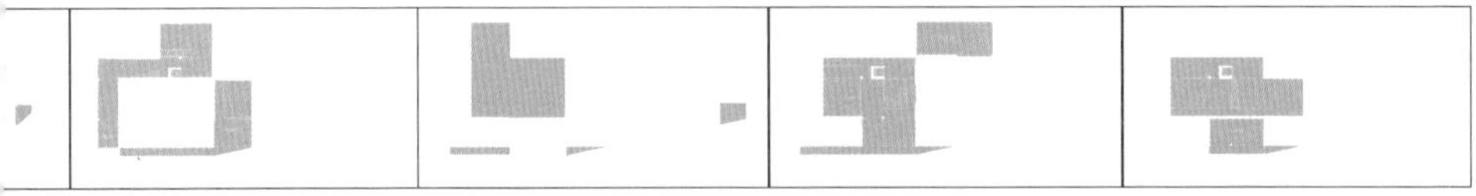

2013年/日本东京/日本桥地区城市更新计划 | 033

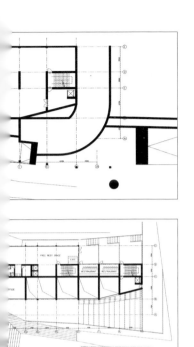

| 七层平面 7F | 八层平面 8F | 九层平面 9F | 十层平面 10F |

JUN. 01
New Year
正月 しょうがつ

MAR. 03
Girl's Day
ひな祭り

MAR. - APR.
Sakura Hanami
花见

MAY. 05
Boy's Day
鯉のぼり

# 花忆 / URBAN GREENING

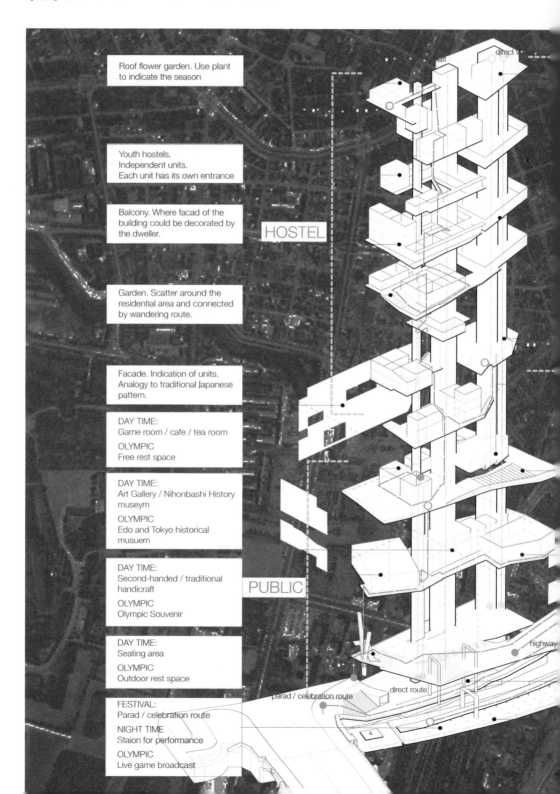

空间构成 SPACE CONSTRUCTION

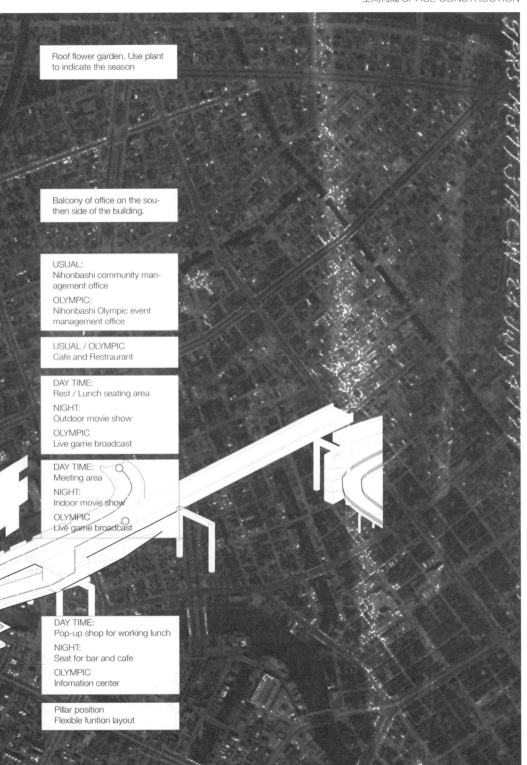

Roof flower garden. Use plant to indicate the season

Balcony of office on the southen side of the building.

USUAL:
Nihonbashi community management office

OLYMPIC:
Nihonbashi Olympic event management office

USUAL / OLYMPIC
Cafe and Restraurant

DAY TIME:
Rest / Lunch seating area
NIGHT:
Outdoor movie show
OLYMPIC
Live game broadcast

DAY TIME:
Meeting area
NIGHT:
Indoor movie show
OLYMPIC
Live game broadcast

DAY TIME:
Pop-up shop for working lunch
NIGHT:
Seat for bar and cafe
OLYMPIC
Infomation center

Pillar position
Flexible funtion layout

2012 年

# 意大利威尼斯 / 大运河滨水建筑更新设计
# RENOVATION OF THE BUILDING IN THE GRAND CANAL VENICE, ITALY

指导教师 程晓青 邹欢 / 学生 陶一兰 钟琳 刘芸 孟宁 孟璠磊 徐妍 缪一新
Teacher: Cheng Xiaoqing, Zou Huan
Student : Tao Yilan, Zhong Lin, Liu Yun, Meng Ning, Meng Fanlei, Xu Yan, Miao Yixin

2012年设计地段位于举世闻名的水城威尼斯中心，大运河边上。设计任务是艺术家居住与工作室，以及相应的展览和餐饮接待功能。在威尼斯设计建筑非常令人激动，同时也非常棘手。多面的城市环境就像意大利人的性格，独特的城市氛围又引发出丰富多彩的创作遐想。如何处理地域、城市、文化、环境等各种因素所提供的限定和线索？如何在堪称经典的威尼斯加入新的时代活力？学生们的答卷令人耳目一新，几个方案分别从建筑原型的角度，从城市生态的角度，从文化延续与创新的角度进行了探索。尤其是"百宝箱"方案，准确地把握了威尼斯欢快旖旎的城市气氛，巧妙地对历史建筑原型加以转译，此时此地，新旧共生。该方案获得2012年大奖。

The site in 2012 was located in the center of Venice, the world famous water-city, just at the side of the Grand Canal. The objective was to design an artist residence and studio with the functions of exhibition and catering reception. It was quite exciting and also very difficult to design a building in Venice. The urban environment was as multifaceted as the character of Italians. The unique urban atmosphere also gave rise to colorful creation reveries. How to handle the limitations and hints provided by multiple factors such as region, city, culture and environment? How to inject new vigor to the classic Venice? The students' answers were refreshing. A few projects respectively made exploration from the perspective of morphology, urban ecology as well as cultural continuation and innovation. In particular, the "Box" project accurately grasped the happy and enchanting urban atmosphere of Venice and skillfully translated the morphology of historic buildings. Now and here, old and new coexist. This project won the grand prize in 2012.

参与院校：
清华大学（中国）
巴黎拉维莱特国立高等建筑学院（法国）
威尼斯建筑大学（意大利）
汉阳大学（韩国）
庆尚大学（韩国）
哈尔滨工业大学（中国）
沈阳建筑大学（中国）

获一等奖院校：清华大学
清华获奖情况：陶一兰荣获一等奖
　　　　　　　钟琳、刘芸、孟宁荣获优秀奖

# 城市百宝箱 / CHEST OF CITY

指导教师 程晓青 邹欢 / 学生 陶一兰 / 2012 年
Teacher: Cheng Xiaoqing, Zou Huan / Student: Tao Yilan / 2012

**教师评语**
设计巧妙地运用当代手法阐释威尼斯的建筑传统，非常准确地反映了大运河与小水巷这两种典型的威尼斯街道空间的性格，建筑简洁明快，轻松流畅，如同维瓦尔第的乐章。

**COMMENTS**
The design interprets the Venetian architectural traditions subtly by contemporary techniques. It accurately reflects the characters of the Grand Canal and small river lane typical urban space of Venice. The building is concise, readable, relaxed and smooth as much as a piece of Vivaldi.

滨水透视 VIEW FROM THE GRAND CANAL

概念生成 CONCEPT GENERATION

# 城市百宝箱 / CHEST OF CITY

TRANSFORMAT

环境分析 ENVIRONMENT ANALYSIS

BIRD

滨水建筑立面
FACADE BY THE
GRAND CANAL

# 城市百宝箱 / CHEST OF CITY

各层平面 FLOOR PLANS

居住单元 LIVING UNITS

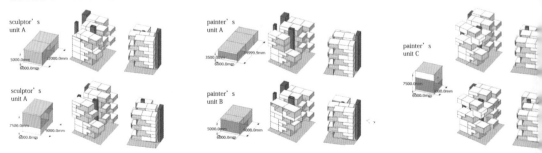

空间分析 SPACE ANALYSIS

2012年/意大利威尼斯/大运河滨水建筑更新设计 | 043

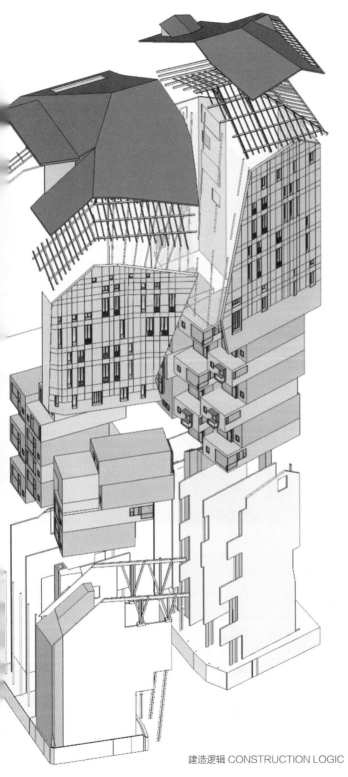

建造逻辑 CONSTRUCTION LOGIC

色彩 COLOURS

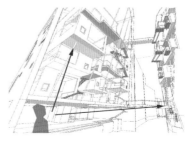

# 城市百宝箱 / CHEST OF CITY

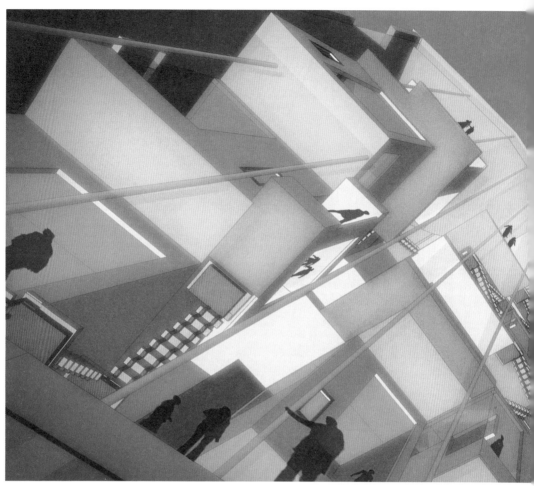

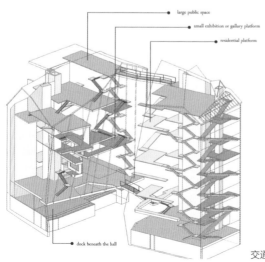

交通分析 STREAMLINE

Dock be

through the glassfloor, people could see

Arge public space

for exhibition, coffee, and dinner . at the first floor or at the top.

2012年/意大利威尼斯/大运河滨水建筑更新设计 | 045

小运河视角透视 VIEW FROM SMALL CANAL

**Mall public platform**

small platforms at the front of artists' studio, so tourists can watch the artists working, and artists can have exchanges with tourists

**Residential platform**

small platforms at the front of artists' homes, facing the small canal, and the exhibition hall opposite. nice view to the canal too, a plat form to show

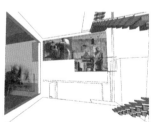

重点空间
MAIN SPACE

# 威尼斯花园 / A GARDEN IN VENICE

指导教师 程晓青 邹欢 / 学生 钟琳 / 2012 年
Teacher: Cheng Xiaoqing, Zou Huan / Student: Zhong Lin / 2012

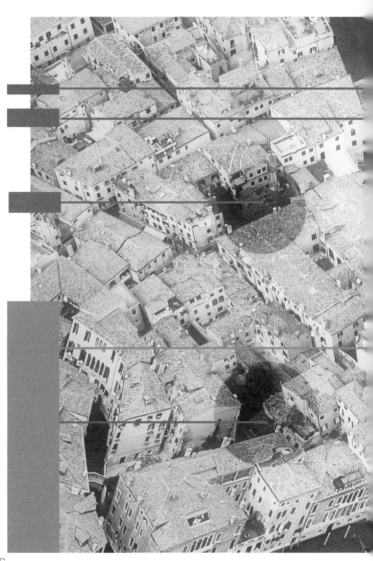

**教师评语**

方案把屋顶花园与垂直绿化引入威尼斯，为水城营造一座立体花园。建筑交通组织与城市公共散步道结合为一体，立面节奏也试图呼应大运河两岸的经典。

**COMMENTS**

The project applies rooftop garden and vertical greening into Venice to create a three-dimensional garden for the waterside town. The building and traffic structure are united with the city's public pedestrain, and rhythm of the façade also attempts to echo with the classic architectural pieces on both banks of the Grand Canal.

## 环境分析 ENVIRONMENT ANALYSIS

| 1500s | 1600s | 1700s | 1800s | |
|---|---|---|---|---|
| Presence of Venitian Garden in Art Woks | Forming of Garden Layout and Design | | | Public Garden Invented for Modern Life | Renaissance of Old G |

The presence of gardens signalled an advanced state of civilization and settlement. All the gardens were owned by each household.

In these elaborate pergolas and benches, people treated gardens as private theaters and had fun there.

Since the land in venece have been becoming more and more valued, many venitian gardens died away.

The venitian gardens could not be more, so they are quite valuable. And the gardens we see today are mostly dated from the 1700s.

Along with the fall of the Republic, the Government decided to create public gardens but they didn't work quite well in the civil life.

When people realized Carlo Scarpa had made

life, they dicided to make the renaissance of old gardens.

### Today: A chance for a new garden

In the site in this design, there is an old garden which gives me inspiration.

This home of artists can be a chance for the combination of public and private life.

### Today: There are not so much green

There are not so much green to be seen in Venice since the irrigation water and land are both valued and most gardens are semi-private.

# 威尼斯花园 / A GARDEN IN VENICE

绿化分析 GREENING

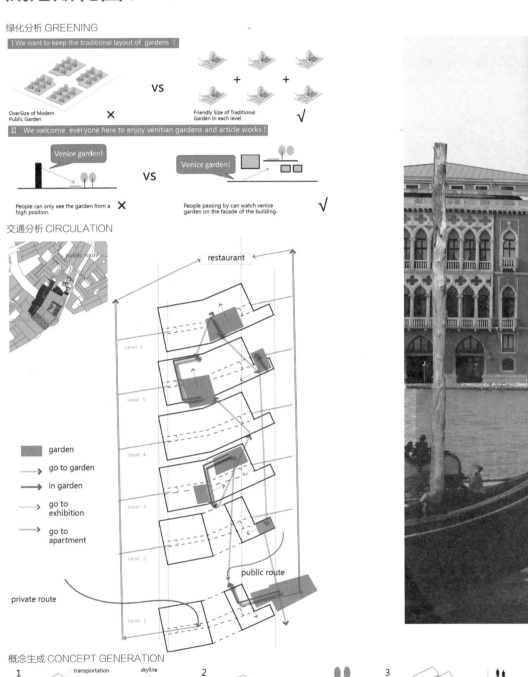

交通分析 CIRCULATION

概念生成 CONCEPT GENERATION

滨水透视 VIEW FROM THE GRAND CANAL

at visitors can
gardens and
he building

add the mode of the Venetian
traditional facade

result4, so that the new building will be in harmony with the environment

# 威尼斯花园 / A GARDEN IN VENICE

绿化构造设计 GREENING DESIGN

垂直绿化 VERTICAL

屋顶绿化 ROOF

表皮绿化 SKIN

0 - entrance
1 - cafe
2 - garden
3 - apartment
4 - art studio
5 - sculptor studio
6 - exhibition room
7 - restaurant
8 - restaurant kitchen

一层 1F
二层 2F
三层 3F
四层 4F
五层 5F
六层 6F
七层 7F

各层平面 FLOOR PLANS

绿化节点 GREENING SPACE

六层花园
GARDEN OF 6F

三层花园
GARDEN OF 3F

五层花园
GARDEN OF 5F

# 缝合威尼斯 / SEWING UP VENICE

指导教师 程晓青 邹欢 / 学生 刘芸 / 2012 年
Teacher: Cheng Xiaoqing, Zou Huan / Student: Liu Yun / 2012

**教师评语**
方案试图以不同城市片段的组合来缝补大运河岸边的空间缺口，各种层级的城市公共空间的引入为建筑带来了活力。

**COMMENTS**
The project tries to make up the spatial gap along the bank of the Grand Canal through the combination of different u fragments. The introduction of urban public space at all le gives a boost of energy to the building.

滨水透视
VIEW FROM THE GRAND CANAL

2012年/意大利威尼斯/大运河滨水建筑更新设计 | 053

概念生成
CONCEPT GENERATION

# 缝合威尼斯 / SEWING UP VENICE

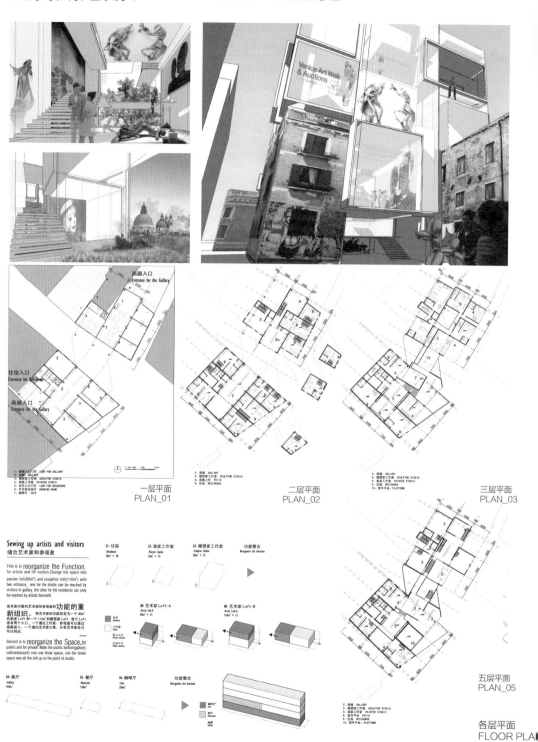

一层平面 PLAN_01

二层平面 PLAN_02

三层平面 PLAN_03

五层平面 PLAN_05

各层平面 FLOOR PLAN

**Sewing up artists and visitors**
缝合艺术家和参观者

First is to **reorganize the Function**, for artists and for visitors. Change the space into painter loft(80m²) and sculptor loft(110m²) with two entrance, one for the studio can be reached by visitors in gallary, the other for the residence can only be reached by artists themself.

首先是对面向艺术家和参观者的**功能的重新组织**。将艺术家的功能改变为一个80m²的画家Loft和一个110m²的雕塑家Loft。每个Loft设有两个入口，一个通过画廊，参观者可以通过画廊进入，一个通过艺术家公寓，只有艺术家自己可以到达。

Second is to **reorganize the Space**, for public and for private. Make the public funtion(gallary, cafe,vestaraunt) into one linear space, use the linear space sew all the loft up on the point of studio.

透视 PERSPECTIVES

# 环形客厅 / LOOP LIVINGROOM

指导教师 程晓青 邹欢 / 学生 孟宁 / 2012 年
Teacher: Cheng Xiaoqing, Zou Huan / Student: Meng Ning / 2012

**教师评语**
连续的建筑公共空间似乎是威尼斯大大小小各种广场的再现，提供了从不同的高度和不同的视角欣赏这个美丽城市的机会。

**COMMENTS**
The successive public spaces of building seem an representation of all squares in Venice. It provide opportunity to appreciate this beautiful city from different he and different perspectives.

## 生成
## CONCEPT GENERATION

**LOOP**
将Loop放于已有建筑之上,建立两岸的联系。
Loop is placed on top of the existing building and the establishment of cross-strait ties.

**SUN**
日照对环境产生了影响,建筑的顶面产生了扭曲,增加了建筑的采光面积。同时,折面屋顶与周围建筑协调。
Sunshine had an impact on the ring, the top surface of the building distortions, increase the area of architectural lighting. At the same time, the fold surface of the roof with the surrounding architecture coordination.

**SCENERY**
通过视线的切割将周边优美的风景收如建筑之中。
Sight cut into the surrounding beautiful scenery income building.

**Living Room**
建筑一层完全打开,形成公共空间,艺术家与当地居民产生积极的互动,形成室外客厅。
The construction of one fully open, the formation of public space. Artists to produce a positive interaction with local residents, the formation of the outdoor living room.

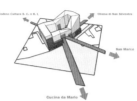

# 环形客厅 / LOOP LIVINGROOM

细部设计概念
DETAIL DESIGN

结构分析
STRUCTURE ANALYSIS

剖面分析
SECETION ANALYSIS

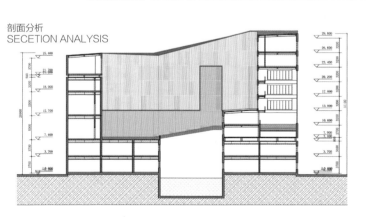

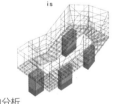

室内分析
INTERIOR ANALYSIS

2012年/意大利威尼斯/大运河滨水建筑更新设计 | 059

一层 1F　　二层 2F　　三层 3F

四层 4F　　五层 5F　　六层 6F

各层平面 FLOOR PLANS

2011年

# 法国奥什 / 军事遗迹改造与再利用
# RENOVATION AND REUSE OF MILITARY HERITAGE, AUCH, FRANCE

指导教师 程晓青 邹欢 / 学生 朱源 顾志琦 刘栎珣 于梦瑶 康茜 符传庆 吴艳珊 吴家东 王舒轶 王焓
Teacher: Cheng Xiaoqing, Zou Huan
Student: Zhu Yuan, Gu Zhiqi, Liu Lixun, Yu Mengyao, Kang Qian, Fu Chuanqing, Wu Yanshan, Wu Jiadong, Wang Shuyi, Wang Han

2011年/法国奥什/军事遗迹改造与再利用 | 061

2011年的题目是思考如何让城市中的飞地优雅地回归，对城市肌理进行有效的缝合。地段位于法国奥什—比利牛斯山区安静的历史小镇，国防部的军产用地回归民用，城市得到新的发展契机。鉴于地段的位置、规模，以及与城市中心的关系，打造新的城市中心的想法被学生们青睐，该想法也得到了地方政府官员的支持。学生方案从城市设计的角度对新中心与老城的接合进行了深入的研究，力图让新旧之间空间流畅顺接，城市生活得以拓展，同时用当代的建筑形式对历史作出回应，留下时代的痕迹。

The project of 2011 aimed to consider how to return the exclave of the city elegantly and complement the urban texture effectively. The site was located in Auch - a tranquil historic town in Pyrenees. The plan was to return the military land of the Ministry of National Defense for civilian use and provide the city with a new development opportunity. Considering the location and scale of the site as well as its relationship with the city center, students favored for the idea of building a new city center. This idea was also supported by local government. From the perspective of city design, students made in-depth research into new center and old city in the design, and tried to achieve the smooth connection of new and old space, so as to expand the city life. Meanwhile, they responded to the history by contemporary building forms, and preserved the historical traces.

参与院校
清华大学（中国）
巴黎拉维莱特国立高等建筑学院（法国）
汉阳大学（韩国）
庆尚大学（韩国）
哈尔滨工业大学（中国）
沈阳建筑大学（中国）

获一等奖院校：清华大学
清华获奖情况：朱源、顾志琦、刘栎珣、于梦瑶、
康茜、符传庆荣获总图设计一等奖
朱源荣获单体设计一等奖

# 奥什复兴计划 / REVIVING AUCH

指导教师 程晓青 邹欢 / 学生 朱源 顾志琦 刘栎珣 于梦瑶 康茜 符传庆 / 2011 年
Teacher: Cheng Xiaoqing, Zou Huan
Student : Zhu Yuan, Gu Zhiqi, Liu Lixun, Yu Mengyao, Kang Qian, Fu Chuanqing / 2011

1. 对于场地内历史年代最久也最为重要的老建筑，我们采取审慎的态度处理。在维掮原有结构基础上，在适当处进行了结构改造，形成异味空间，以满足住宅办公个性空间之外的公共交流活动。老建筑中新空间的加入同样给与使用者以奇妙的感受；
2. 新建筑的加入，一方面丰富了公共活动的功能，另一方面也对场地进行了划分，在南北侧分别形成了内街和广场，以此将右岸街道在空间和视觉上延续开来。新建筑面向河岸的端部底部架空，通过场地的台阶高差处理，将河岸顺势过渡到广场和建筑入口，吸引人群。

For the most ancient and important building on the site, we choose to reform it with a careful attitude. When preserving the original structure, we transform the structure at certain places to form some disparate spaces, in order to meet the communicational needs bedsides the ordinary business activity and soho life. The adding of the spaces also bring a fantasstic feeling to the people comparing to the existing old structure. The construcion of the new building, on the one hand riches the public activity, on the other hand divides the site into a commericial street and a square. The street can have visual connection with the street on the east bank. The side of the new building facing the river bank was structed above the land. The steps which connect the bank level and the site level attracts people from the river bank to the buildings.

## 教师评语
城市设计在尊重历史建筑和激活沉睡街区之间找到了较好的平衡，以一种积极主动的方式巧妙地把河两岸的建筑空间联结起来，并形成新兴街区的活力。建筑单体之间和而不同，相映成趣，形成丰富又整体的空间环境。

## COMMENTS
The urban design interprets a better balance between respecting the historical buildings and activating the sleepy urban blocks. It integrates the urban spaces on both banks of the river subtly by a positive and initiative manner, and forms the vitality of emerging urban blocks. The united but different building units form delightful contrast and also create a colorful and integrated space environment.

环境分析 ENVIRONMENT ANALYSIS

地段区位 / Site Location  交通路网 / Traffic Network  开放空间 / Open Space  建筑现状 / Building Status

2011年/法国奥什/军事遗迹改造与再利用

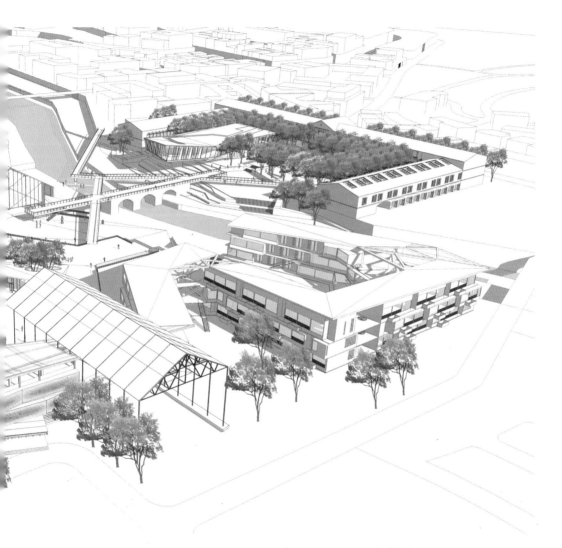

概念生成 CONCEPT GENERATION

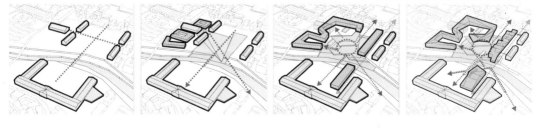

# 奥什复兴计划 / REVIVING AUCH

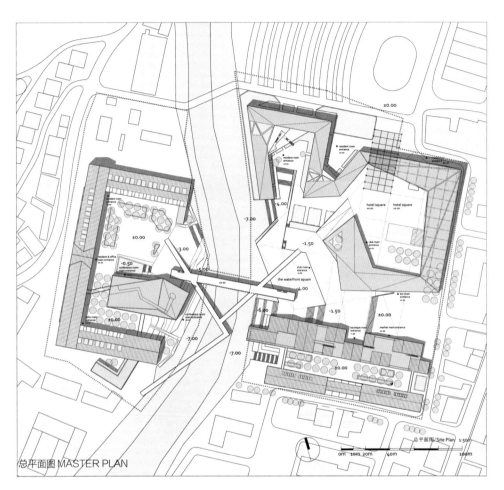

总平面图 MASTER PLAN

综合分析 ANALYSIS

高差示意图 /ALTITUDE DIAGRAM   历史元素 /HISTORY ELEMENTS   功能示意图 /FUNCTION DIAGRAM   流线示意图 /FLOW LINES

北侧透视图 VIEW FROM NORTH   南侧透视图 VIEW FROM SOOTH

## 北侧场地现状 STATUS OF NORTHERN PART

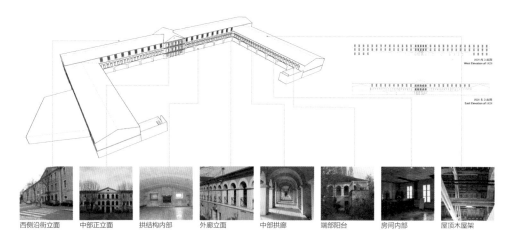

西侧沿街立面　中部正立面　拱结构内部　外廊立面　中部拱廊　端部阳台　房间内部　屋顶木屋架

## 旧建筑改造示意 RENOVATION OF OLD BUILDING

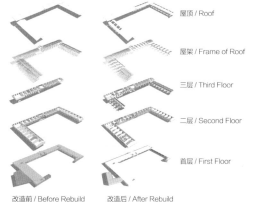

屋顶 / Roof
屋架 / Frame of Roof
三层 / Third Floor
二层 / Second Floor
首层 / First Floor

改造前 / Before Rebuild　　改造后 / After Rebuild

侧段屋顶阳台 / 中段三层阳台

中部异质空间 / 首层露出拱结构

南侧一层建筑改造成院 / 西立面的加建遮阳措施

## 北侧建筑鸟瞰 BIRD'S EYE VIEW OF NORTHERN PART

# 奥什复兴计划 / REVIVING AUCH

北侧建筑改造 RENOVATION OF NORTHERN PART

1828 北段立面图（场地内）/NORTH ELEVATION OF 1828

东立面图 /EAST ELEVATION

北立面图 / NORTH ELEVATION

西立面图 / WEST ELEVATION

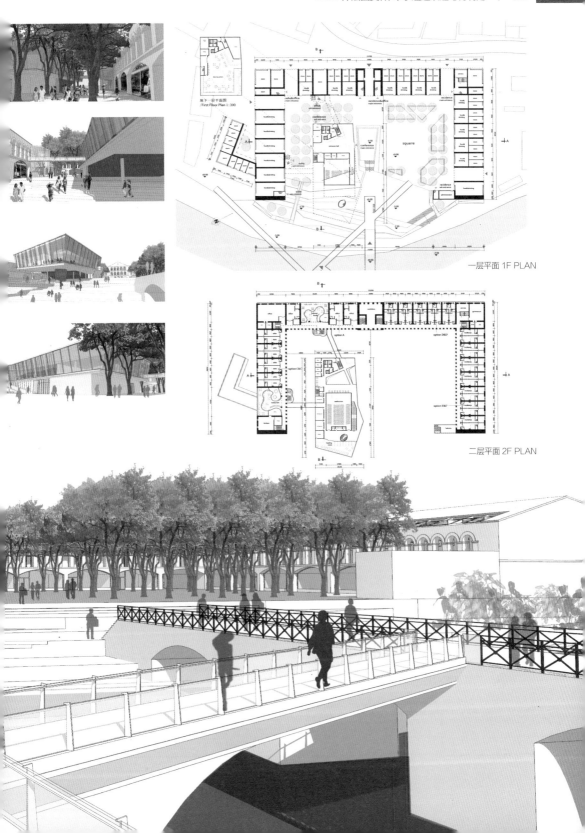

068 | 国际建筑工坊 AIAC

# 奥什复兴计划 / REVIVING AUCH

南侧建筑改造 1 RENOVATION OF SOUTHERN PART 1

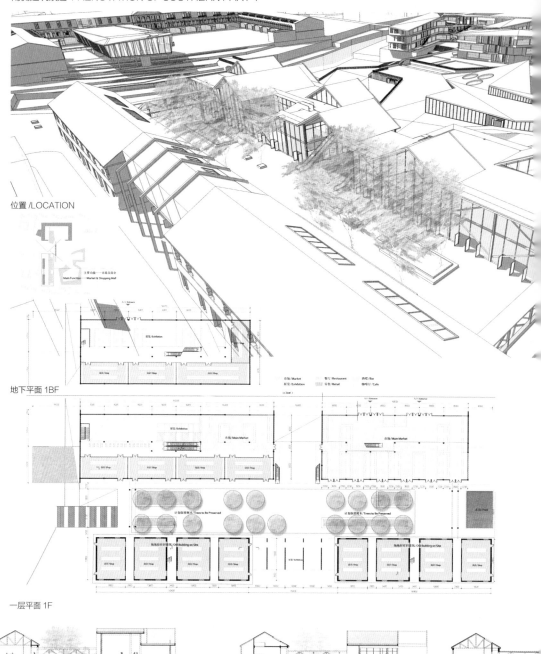

2011年/法国奥什/军事遗迹改造与再利用 | 069

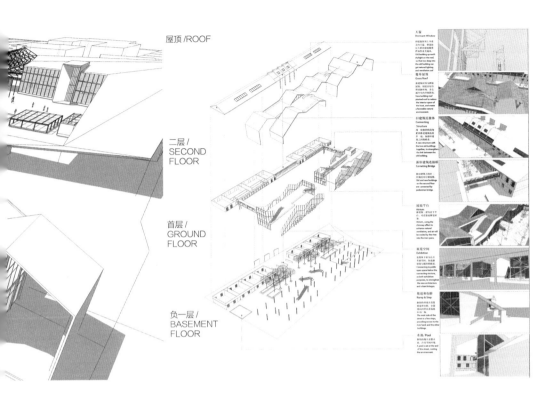

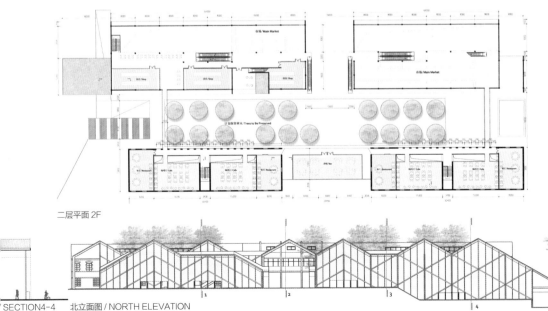

二层平面 2F

SECTION4-4  北立面图 / NORTH ELEVATION

# 奥什复兴计划 / REVIVING AUCH

南侧建筑改造 2 RENOVATION OF SOUTHEAN SITE 2

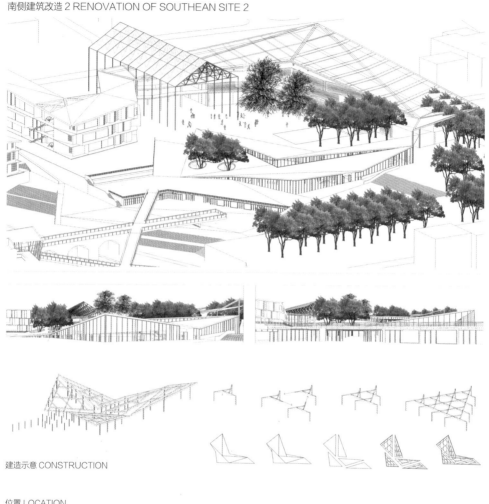

建造示意 CONSTRUCTION

位置 LOCATION

节点详图 DET

剖面图 SECTI

2011年/法国奥什/军事遗迹改造与再利用 | 071

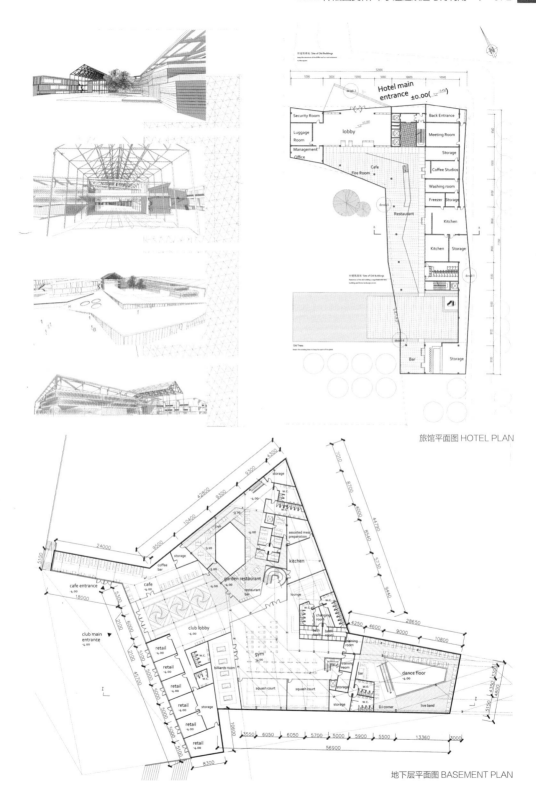

旅馆平面图 HOTEL PLAN

地下层平面图 BASEMENT PLAN

# 奥什复兴计划 / REVIVING AUCH

南侧建筑改造 3 RENOVATION OF SOUTHEAN PART 3

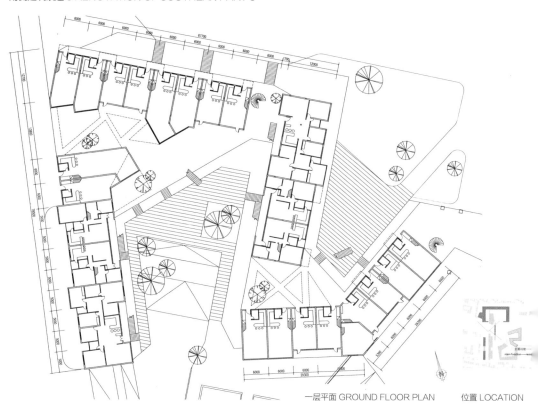

一层平面 GROUND FLOOR PLAN    位置 LOCATION

住宅内景 COURT OF RESIDENCE

2011年/法国奥什/军事遗迹改造与再利用 | 073

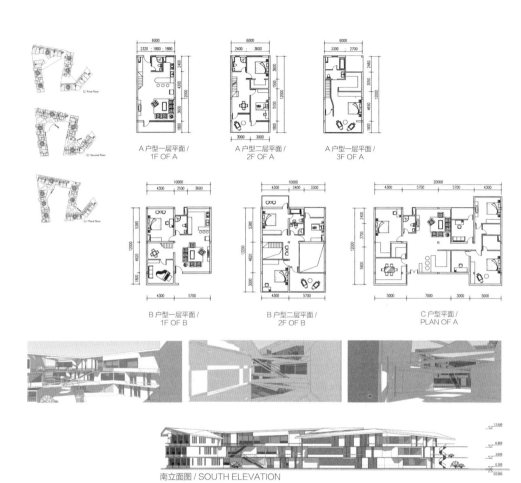

A 户型一层平面 / 1F OF A

A 户型二层平面 / 2F OF A

A 户型三层平面 / 3F OF A

B 户型一层平面 / 1F OF B

B 户型二层平面 / 2F OF B

C 户型平面 / PLAN OF A

南立面图 / SOUTH ELEVATION

东立面图 / EAST ELEVATION

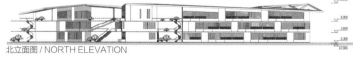

北立面图 / NORTH ELEVATION

西立面图 / NORTH ELEVATION

2010年

# 法国巴约纳 / 滨水空间更新计划
# RENEWAL OF RIVER BANK, BAYONNE, FRANCE

指导教师 程晓青 邹欢 / 学生 李昆 齐际 刘伯宇 秦岭 吴锡嘉 任凯 王禹 唐任杰
Teacher: Cheng Xiaoqing, Zou Huan
Student : Li Kun, Qi Ji, Liu Boyu, Qin Ling, Wu Xijia, Ren Kai, Wang Yu, Tang Renjie

2010年/法国巴约纳/滨水空间更新计划 | 075

位于巴斯克地区的 Bayonne 物产丰富，深厚的历史文化底蕴更是其引以为豪的财富。随着社会与经济的转型，原有的军队占地将归还给城市，如何将这块飞地还归城市，尤其是如何将其融入 Bayonne 城市肌理中去，使城市丰富多彩的生活得以延续，是 2010 年的主题。地段位于巴斯克地区的母亲河——拉杜尔河畔，同时紧邻的山丘上有一座著名的历史建筑 Bayonne 城堡，火车站也近在咫尺。复杂的地段特征，城市历史文化的考量，交通的组织，景观的需要，2010 年的题目涵盖城市和建筑更新，对于东方的学生是一个艰巨的考验。得奖方案充分考虑了交通和景观的呼应，把"蒙太奇"和"步移景异"植入到建筑体验中，创新而浪漫的建筑形式也给人带来耳目一新的感觉。

Located in the Basque region, Bayonne is abundant in resources. The rich history and culture are the wealth that Bayonne takes pride in. With the transformation of social economy, original military occupied land will be returned to the city. For this reason, the theme in 2010 was to integrate it into the urban texture of Bayonne and continue the colorful urban life. The site was located at the river bank of Latur - the mother river of Basque. Meanwhile, there was a famous historical building - Bayonne Castle adjacent to the hill. The railway station was also near at hand. With complicated site characteristics, consideration to urban history and cultures, organization of traffic and needs of landscape, the project in 2010 was a great test for the oriental students as it covered both the city and the buildings. The winning project gave full consideration to the response between traffic and landscape, implanted the "montage" and "scenery changing with walking" into the building experience, and brought a new perspective for people by virtue of innovative and romantic building form.

参与院校                                    获一等奖院校：清华大学
清华大学（中国）                            清华获奖情况：李昆、齐际荣获一等奖
巴黎拉维莱特国立高等建筑学院（法国）                     刘伯宇、秦岭荣获二等获
汉阳大学（韩国）
庆尚大学（韩国）
哈尔滨工业大学（中国）
沈阳建筑大学（中国）

# 城市取景框 / VIEW FRAME OF THE CITY

指导教师 程晓青 邹欢 / 学生 李昆 齐际 /2010 年
Teacher: Cheng Xiaoqing, Zou Huan / Student: Li Kun, Qi Ji / 2010

**教师评语**

把连在一起的建筑切断，通过断开的空隙去观赏城市，这种景框式的做法因为地段旁边有一条铁路通过而变得更加有趣，由此带来的建筑形体的变化也就顺理成章，而为此对结构和构造的研究使方案更为深入。

**COMMENTS**

By the discontinuity of connected buildings, the design enables people to appreciate the city through disconnected space. This practice of view box is more interesting against a railway near the site, so the resulting change in the architectural form is well-reasoned and thus facilitate the in-depth research into the construction and structure.

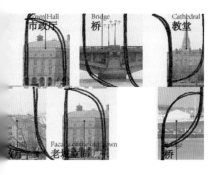

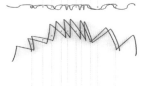

VACUMM BETWEEN CURVES
曲线间的缝隙

概念生成 CONCEPT GENERATION

SURFACES BETWEEN VACUMM
间隙间的曲面

VACUMM BETWEEN CURVES
曲线间的缝隙

SURFACES BETWEEN VACUMM
间隙间的曲面

# 城市取景框 / VIEW FRAME OF CITY

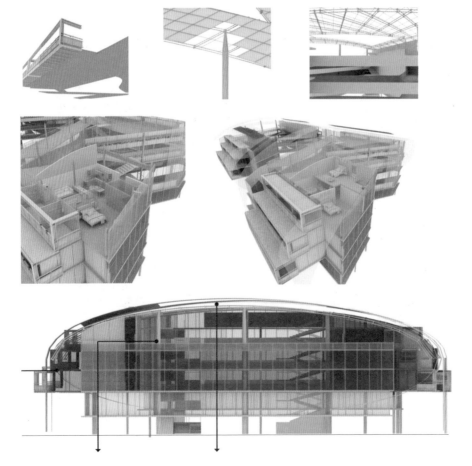

细部设计 DETA

剖面设计 SECTI

2010年/法国巴约纳/滨水空间更新计划 | 079

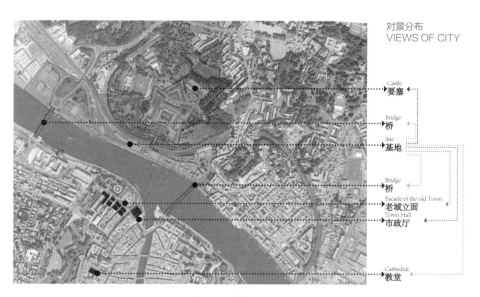

对景分布
VIEWS OF CITY

Castle 要塞
Bridge 桥
Site 基地
Bridge 桥
Facade of the old Town 老城立面
Town Hall 市政厅
Cathedral 教堂

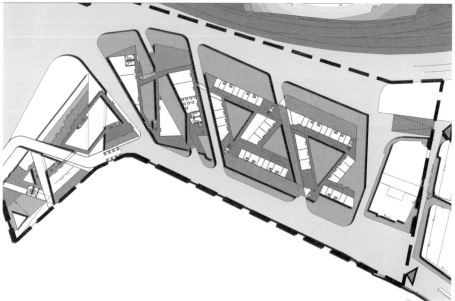

平面设计 FLOOR PLAN

# 活力极 & 舒适岛 / LIVELY POLE & COZY ISLAND

指导教师 程晓青 邹欢 / 学生 刘伯宇 秦岭 / 2010 年
Teacher: Cheng Xiaoqing, Zou Huan / Student: Liu Boyu, Qin Ling / 2010

**教师评语**
打破住宅单调的形象和把绿化引入建筑组群是这个方案的出发点，为此公建化的建筑处理和屋顶花园结合，沿河岸的城市公共空间与住宅相互交错，增强城市生活的互动性。

**COMMENTS**
The basic idea of this project is to break down the tedious image of residential buildings and introduce greening into building complexes. Hence, the interactive city life is strengthened by the combination of public architectural treatment and rooftop garden, so will mutual interlacing between urban public space and residential building along the river bank be of help.

概念生成 CONCEPT GENERATION

2010年/法国巴约纳/滨水空间更新计划

# 活力极 & 舒适岛 / LIVELY POLE & COZY ISLAND

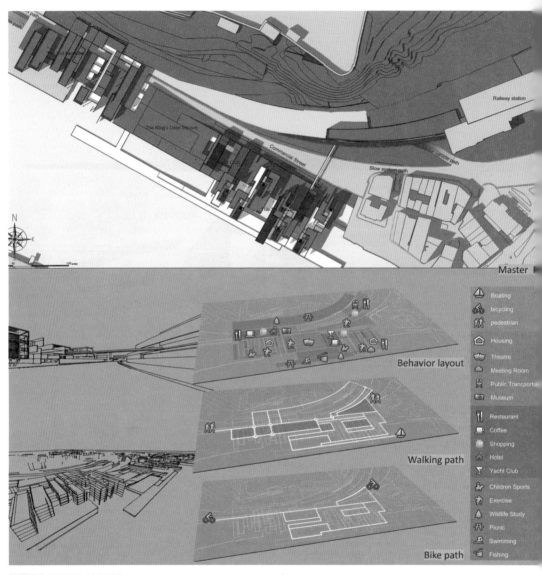

总图设计 SITE PLANNING

剖面设计 SECTION DESIGN

2010年/法国巴约纳/滨水空间更新计划 | 083

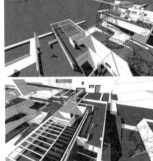

Standard layer Plane  Unit Distribution

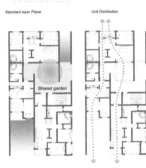

The external environment  Wind Road

住宅单元 RESIDENCE UNIT

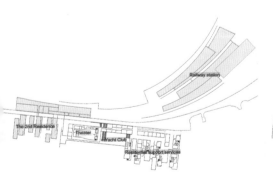

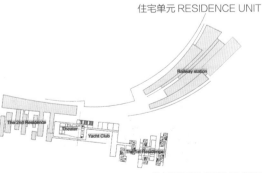

各层平面 FLOOR PLANS

2009年

# 韩国晋州 / 泗川滨海现代音乐厅
# MODERN CONCERT HALL BY SEA IN SICHUAN, JINZHOU, R.O. KOREA

指导教师 栗德祥 程晓青 邹欢 / 学生 戴天行 闫晋波 曲直 金世中 苏海星 赵瞳 孙小暖 揭小凤
Teacher: Li Dexiang, Cheng Xiaoqing, Zou Huan
Student: Dai Tianxing, Yan Jinbo, Qu Zhi, Kim SeJung, Su Haixing, Zhao Tong, Sun Xiaonuan, Jie Xiaofeng

2009年的题目在韩国晋州的海边,任务是设计一个音乐厅。这里远离喧嚣的城市,依山傍水,视线开阔。建筑与自然的交流和对话是学生们关注的主题,同时紧张的地段内丰富的地形变化也为设计带来挑战。学生的设计方案探索了建筑与风景的关系,建筑借景、造景与人在建筑环境中感受景观的体验,同时对建筑垂直空间的流动与贯通作了大胆的尝试。

The project in 2009 was to design a music hall onthe seaside of Chinju, South Korea. This place was far away from the noisy city and situated between hills and waters, with an open and wide view. The theme that students concerned was to achieve the exchange and dialogue between the buildings and the nature. Meanwhile, the rich topographic changes in this tense site also challenged the design itself. In the projects, students explored the relationship between buildings and scenery, and created the experience of borrowing spaces and landscap to make people feel the landscape in the building environment. Meanwhile, they made a bold attempt to the mobility and linkage of vertical building space.

参与院校　　　　　　　　　　　　　　获一等奖院校:巴黎拉维莱特国立高等建筑学院
清华大学(中国)
巴黎拉维莱特国立高等建筑学院(法国)
汉阳大学(韩国)
庆尚大学(韩国)
罗马大学(意大利)

# 梯田 / TERRACE

指导教师 栗德祥 程晓青 邹欢 / 学生 戴天行 闫晋波 / 2009 年
Teacher: Li Dexiang, Cheng Xiaoqing, Zou Huan / Student: Dai Tianxing, Yan Jinbo / 2009

**教师评语**
建筑似乎是自然山丘和梯田的延续,地形高差正好成为连贯流动的公共空间的分隔,从室外到室内,一气呵成。

**COMMENTS**
The building seems to be the continuation of natural hills and terrace. The elevation difference in terrain naturally becomes a separation of the continuously flowing public space, so the sense of entirety is manifested from outdoor to indoor.

环境分析 ENVIRONMENT ANALYSIS

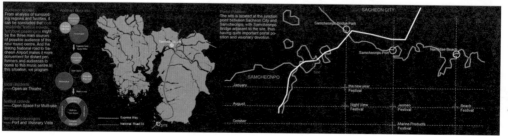

2009年/韩国晋州/泗川滨海现代音乐厅 | 087

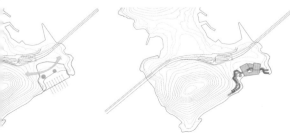

# 梯田 / TERRACE

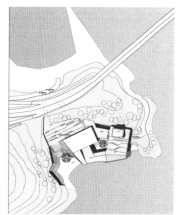

总平面 SITE PLAN

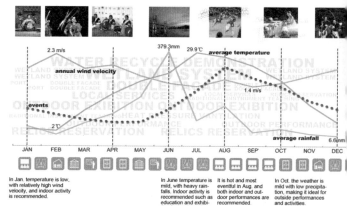

生态要素 ECO ELEMENTS

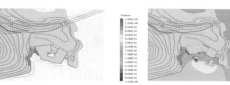
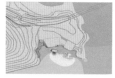
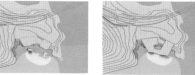

环境分析 ENVIOREMENT DIAGRAM

体型构成 TRANSFORMATION DIAGRAM

交通构成 CIRCULATION DIAGRAM

立面 FACADE

2009年/韩国晋州/泗川滨海现代音乐厅 | 089

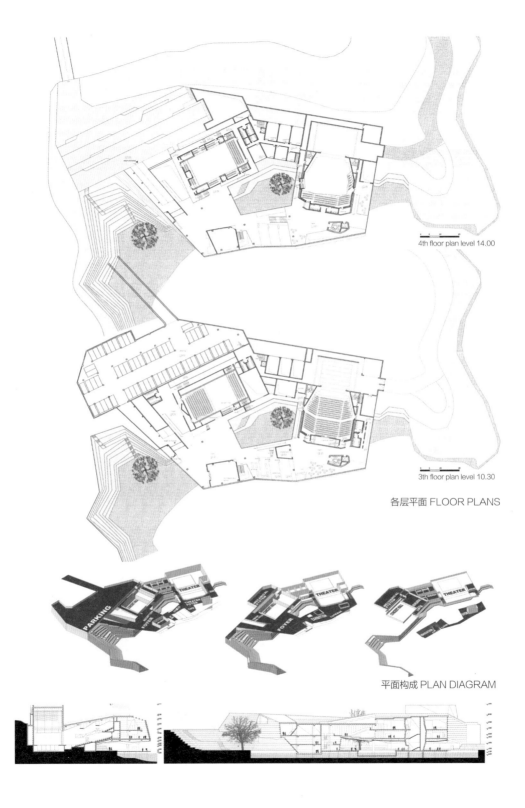

4th floor plan level 14.00

3th floor plan level 10.30

各层平面 FLOOR PLANS

平面构成 PLAN DIAGRAM

# 梯田 / TERRACE

室内空间 INSIDE EVENT SPACE

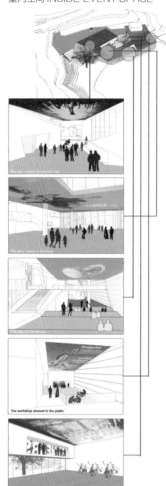
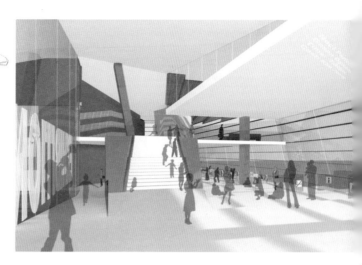

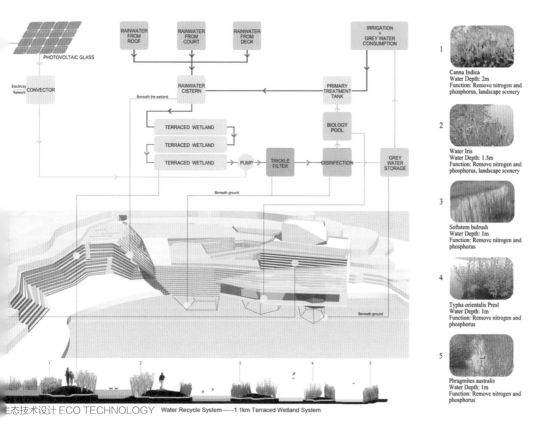

生态技术设计 ECO TECHNOLOGY    Water Recycle System------1.1km Terraced Wetland System

1 Canna Indica
Water Depth: 2m
Function: Remove nitrogen and phosphorus, landscape scenery

2 Water Iris
Water Depth: 1.5m
Function: Remove nitrogen and phosphorus, landscape scenery

3 Softstem bulrush
Water Depth: 1m
Function: Remove nitrogen and phosphorus

4 Typha orientalis Presl
Water Depth: 1m
Function: Remove nitrogen and phosphorus

5 Phragmites australis
Water Depth: 1m
Function: Remove nitrogen and phosphorus

室外空间 EXTERIOR SPACE

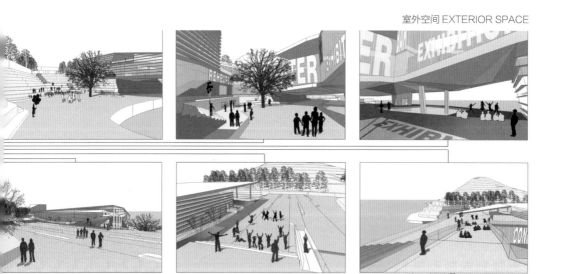

2008年
# 中国深圳 / 蛇口区滨海创意中心
# CREATIVE CENTRE BY SEA IN SHEKOU SHENZHEN, CHINA

指导教师 栗德祥 程晓青 邹欢
学生 熊星 秦嘉煜 李烨 陈忱 乔会卿 都文娟 郑娟 孔媛 林展鹏 吴焕 马晓瑛 傅平川 赵鹏
Teacher: Li Dexiang, Cheng Xiaoqing, Zou Huan
Student: Xiong Xing, Qin Jiayu, Li Ye, Chen Chen, Qiao Huiqing, Du Wenjuan, Zheng Juan, Kong Yuan, Lin Zhanpeng, Wu Huan, Ma Xiaoying, Fu Pingchuan, Zhao Peng

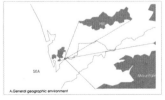

A. General geographic environment

B. The important status in the urban scale

C. The arrangement of current coastline

2008年/中国深圳/蛇口区滨海创意中心 | 093

奥运前夕，"国际建筑工坊"又来到中国。这一次我们选择了深圳，一个全新的城市，一个雄心勃勃的城市。如何在一个缺乏历史积淀的城市设计建筑？如何处理超大尺度建筑与城市环境的关系？如何让城市标志性建筑更具有人性活力？如何在大型公共建筑中运用生态技术？学生的方案对这些问题进行了探讨。从城市公共空间入手解决问题是积极的应对方法。建筑不论大小，只有与城市空间紧密结合、融合渗透，才能相得益彰、持续永恒。

Right before the Olympic Games, the AIAC came to China again. This time we selected Shenzhen, a new and ambitious city. How to design buildings in such a city lacking historical accumulation? How to handle the relationship between super-large buildings and urban environment? How did they make the city's landmark building more vibrant and humane? How to apply ecological technologies in the large-scale public buildings? Students discussed these topics in their projects. A positive measure was to solve problems from the perspective of urban public space. Regardless of the size of the building, it could shine more brilliantly and remain perpetual only by tight combination and integration with urban space.

| 参与院校 | 获一等奖院校：清华大学 |
| --- | --- |
| 清华大学（中国） | 清华获奖情况：熊星、秦嘉煜荣获一等奖 |
| 巴黎拉维莱特国立高等建筑学院（法国） | 李烨、陈忱荣获二等奖 |
| 汉阳大学（韩国） | |
| 庆尚大学（韩国） | |
| 深圳大学（中国） | |

# 垂挂城市 / HANGING DOWNTOWN

指导教师 栗德祥 程晓青 邹欢 / 学生 熊星 秦嘉煜 / 2008 年
Teacher: Li Dexiang, Cheng Xiaoqing, Zou Huan / Student: Xiong Xing, Qin Jiayu / 2008

## 教师评语

建筑似乎是自然山丘的延续，用巨构来回应深圳—蛇口—海边这个非常特别的地段，创意大胆，具有想象力。

## COMMENTS

The project responds with such a special site in Shenzhen-Shekou-Seaside by a mega structure, with bold creativity and imagination.

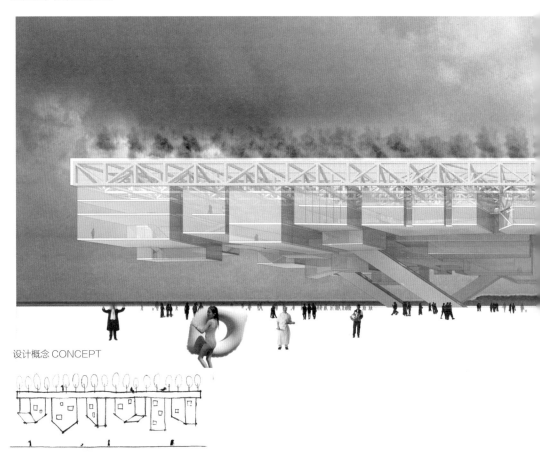

### 设计概念 CONCEPT

### 环境分析与应对 ENVIRONMENT ANALYSIS

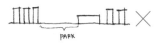

**Limited park space**
The nearby park serves as an important public space for both residents and tourists. According to the current planning, the park is so narrow, thus if a solid building stands there, it will limited the space area for public entertainment.

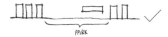

**Expanded park space**
By leaving the ground empty, the of the park is expanded. It is a e way to enhance the public acti maximize the use of public land

**Blocked the sea sight**
The current planning makes the littoral space filled by almost residences with rare public open spaces. Thus the sea, as the lanscape resource of city, seems more precious. It is unfair for any building to block the view of sea.

**Opened the sea sight**
For the importance of the sea sacpe resource. It is actually a the megastructure to make the sight accessible for the public t say the road and plaza, rather to block there or arrogate selfish

2008年/中国深圳/蛇口区滨海创意中心 | 095

= S

without any construction
ilable for people's
s, is defined intitially
value of S.

→

Area = 0

While A typical buildings covers the site where it is located, the public space is blocked. Thus there is no open area leave to the public activities.

Area = S/2

There is a fashion of folding the roof into the ground and finally formed a slope. However, people would less like to stay on such non flat place, then its utility takes discount.

Area = S

Another pragmatic strategy is to make the roof a garden or plaza available for public, that will completely offset the total area it covers.

Area = 2 S

The most efficient solution will be this one: to leave ground maintain public feature, at the same time use the roof as a garden that also open to public. The space will be doubled.

# 垂挂城市 / HANGING DOWNTOWN

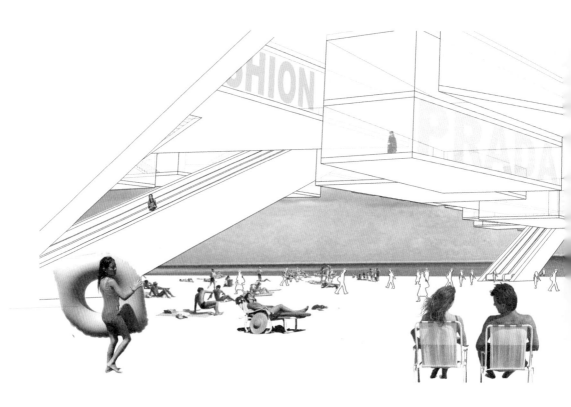

## 技术策略 TECHNICAL STRATEGY

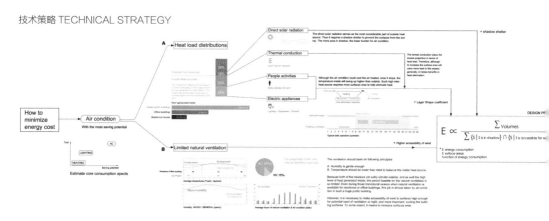

2008年/中国深圳/蛇口区滨海创意中心 | 097

公共空间 PUBLIC SPACE

house

Cinema

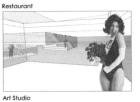
Restaurant

Pedestrian street 1

strian street 2

Art Studio

Art Studio

Roof Garden

# 垂挂城市 /HANGING DOWNTOWN

## 技术设计 TECHNICAL DESIGN

 electricity wire

**Electricity wire net 10m x 10m**
The electricity net is contributed into a 10m x 10m density webs and covers the roof evenly. They converges into five main wires and then are connected with the city electricity net. Meanwhile the internet wires are associated with the electricity wires.

■ machine room

**Air-conditioner pipes**
The air-conditioning system here adopted a central type. The wind outlet are scattered on the roof plane in a pattern of 10m x 15m rectangle. And the machine room is placed in the basement.

**Stress calculation in SAP2000**

**Stress distribution**

**Component design**

 Brace under...

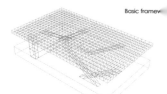 Basic framew...

fresh water supply
sewege outlet

**Fresh water supply and sewege**
The freshe water system is connected with the city water supply system and brought up in five main pipes. Then they are distributed into a 20m-density paralles on the roof. The sewege produced by each unit is pumped up by a small solar-energy drived pump to the main pipes on the roof structure.

● manure station

**Treated manure for the green roof**
Five public lavatories are designed near each escalator for people. Their wastes are treated briefly, discomposted off the NH3. Then they are pumped up and circled around to nurture the plantes on roof.

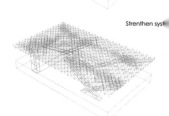 Strenthen syst...

■ water treatment spot

**Roof drainage outlet**
The rain water collected in the roof are drained out by 25 drainage outlet.Still, some of it is kept and treated right on the mini water treatment station, which hiden in the roof structure, and circled back to water the plants on the roof.

■ commercal unit   ● escape holes

**Fire emergency escape track**
For the fire emergency track,we adopte the conventional way used in Japanese residence. The fire emergency hole appears on each sigle unit some even have two or three, depending on the scale of the unit with a ripe ladder which can reach the ground.

## 各层平面 FLOOR PLANS

**Basement-1**
Altitude = -6m

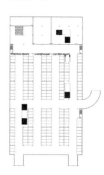

**Basement-1**
Altitude = -12m

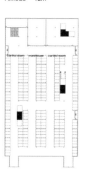

**Roof garden plan**

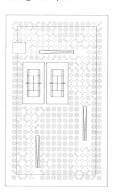

**Tree & Structurelines**

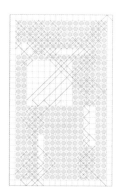

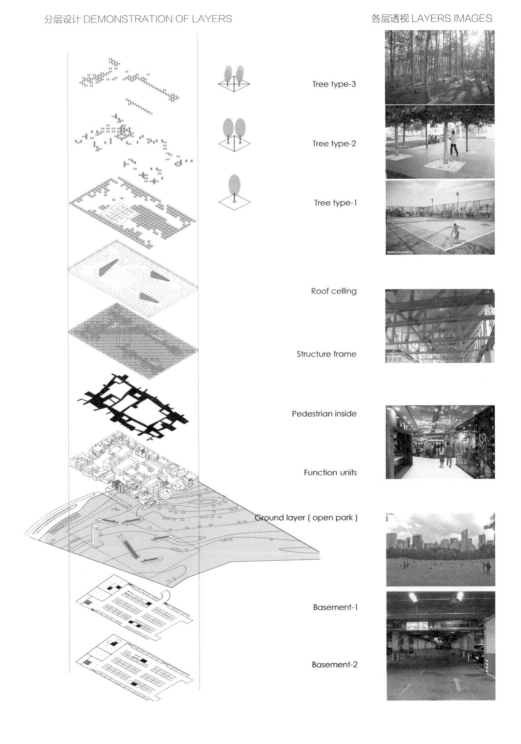

# 垂挂城市 / HANGING DOWNTOWN

各层平面 FLOOR PLANS

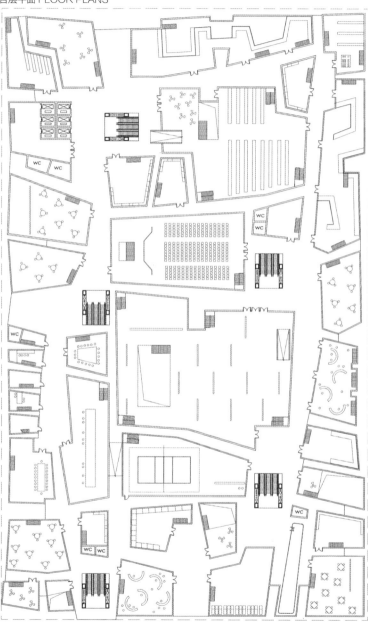

Floor -1  Altitude =18m

Floor -1  Altitude=

商业单元 COMMERCIAL

cafeteria + book shop

architectesl studio + gallery

car selling + exhibition

body building gym

立面 FACADE

South elevation

2008年/中国深圳/蛇口区滨海创意中心

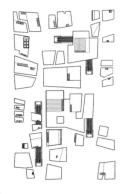

Floor -2 Altitude=16m

Floor -3 Altitude=13m

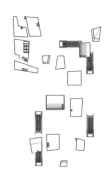

Floor -4 Altitude=10m

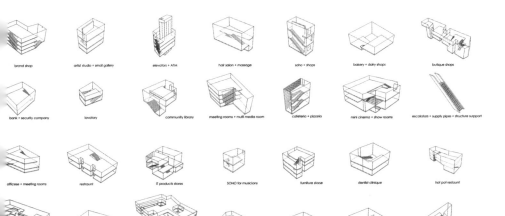

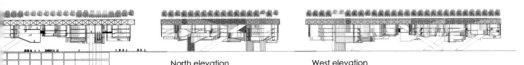

North elevation

West elevation

# 风之翼 / WING OF WIND

指导教师 栗德祥 程晓青 邹欢 / 学生 李烨 陈忱 / 2008 年
Teacher: Li Dexiang, Cheng Xiaoqing, Zou Huan / Student: Li Ye, Chen Chen / 2008

### 教师评语
把机翼与气流原理运用到建筑上来，解决巨构建筑的通风与采光，并由此产生建筑形式，不失为有趣的探索。

### COMMENTS
The project applies the theory of wing and air flow on the building, which not only solves the ventilation and daylight issues of the mega structure but also gives a form. So, it can yet be regarded as an interesting exploration.

## 概念生成 CONCEPT GENERATION

**Mega-terminal**
Based on the very site and its terminal position on the urban landscape axis, our strategy is to establish a Mega-complex as a new landmark along the seashore for the urban space and maximize the inner efficiency for the complex itself. We also put up the concept of "interior Time-Zone" to rethink the program layout of a commercial complex.

**Vertical Sectional Ventilation**
According to the official climate data of SHEKOU, SHENZHEN and the inspiration of wing section design, we propose the strategy of "Vertical Sectional Ventilation" to solve the mega-structure's ventilation problem and raise a consistency strategy for the changing environment in the whole year of SHENZHEN.

**Tubes In and Atriums Out**
We put forward different modes of ventilation, simulated them in CFD soft and finally optimized a system of vertical ventilation with well collaborating inlet tubes and outlet atriums. Sectional details were also carefully designed.

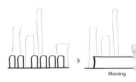

# 风之翼 / WING OF WIND

## 环境分析 ENVIRONMENT ANALYSIS

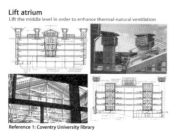

Lift atrium
Lift the middle level in order to enhance thermal-natural ventilation

Reference 1: Coventry University library

Wind deflector
adaptable for wind from all directions

Reference 2: BEDZED

Wind tube as structure
New air goes to different floors separately via structure

Reference 3: Sendai Mediatheque

| | TEMP. | W.S. | HUM. |
|---|---|---|---|
| ① High temperature (jun to sep - diurnal) | ⊗ high | | |
| ② Low temperature (jan and feb) | ⊗ low | | |
| ③ Eligible wind speed, both temperature and humidity are fine | ✓ | ✓ | ✓ |
| ④ Eligible wind speed, fine temperature, bad humidity | ✓ | ✓ | ⊗ |
| ⑤ Low wind speed, both temperature and humidity are fine | ✓ | none | ✓ |
| ⑥ Low wind speed, fine temperature, bad humidity | ✓ | none | ⊗ |
| ⑦ High wind speed, storm or typhoon | | typhoon | |

✓ temperature: 16℃~28℃   wind speed: 3m/s~5m/s   humidity: 30%~70

## 空间生成 TRANSFORMATION

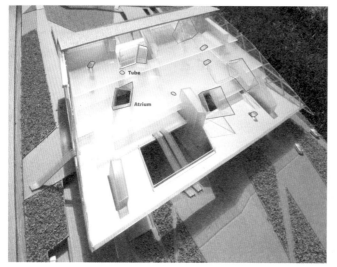

## 通风设计与模拟
## VENTILATION SIMULATION

After carefully comparing and testing different ventilation modes, we select the most crucial ecology factors on this site to utilize as part of our concept —— the most frequent sea wind in transition seasons. We approach the shape and direction to maximize the power of sea wind, and value of seascape.

Also, in order to make the architecture more adaptable to the environment, we design the rotatable wind deflectors to let in wind from all directions, and atriums to get thermal-natural ventilation when without wind.

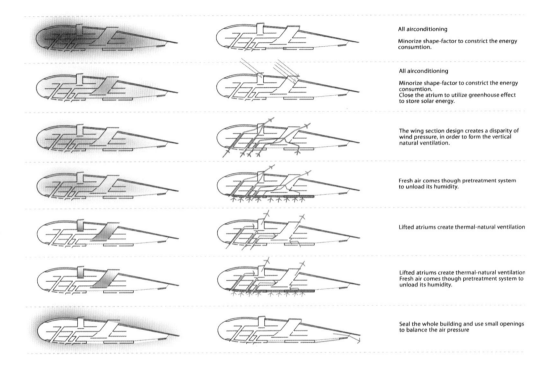

All airconditioning
Minorize shape-factor to constrict the energy consumtion.

All airconditioning
Minorize shape-factor to constrict the energy consumtion.
Close the atrium to utilize greenhouse effect to store solar energy.

The wing section design creates a disparity of wind pressure, in order to form the vertical natural ventilation.

Fresh air comes though pretreatment system to unload its humidity.

Lifted atriums create thermal-natural ventilation

Lifted atriums create thermal-natural ventilation
Fresh air comes though pretreatment system to unload its humidity.

Seal the whole building and use small openings to balance the air pressure

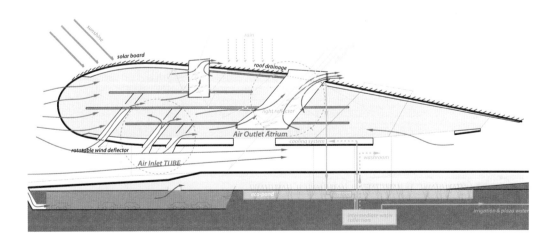

# 风之翼 / WING OF WIND

## 风的利用 USE OF WIND

### The Wing

**Vertical Ventilation Strategy**

As a commercial complex, mega-structure building offers high running efficiency, yet meanwhile brings along a problem of ventilation. We apply a wing section to the mega-envelope, solving the problem with Vertical Ventilation Strategy.

Our Solution

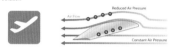

Get ventilation power from the lift force of a wing section.

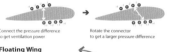

**Floating Wing**

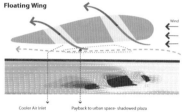

### The Wind

**Natural Ventilation in Transition Seasons**

In SHENZHEN, natural ventilation in transition seasons could give great merits to energy conservation, while in summer and winter the building must seal itself to suffer minimum energy loss. Precise definition of transition seasons for the very location and careful choice of wind direction is highly demanded.

**Step1** - Transition Seasons Screening

We carefully screening Transition Seasons particularly for the site according to for how many hours the temperature is exactly between 16C-28C [The Thermal Comfort Temperature]. Parameters include both ShenZhen's Climate and the site's Microclimate.

**Step2** - Wind Direction Choice

Two Wind Directions are chosen at this phase according to wind frequency statistics data during the Transition Seasons.

**Step3** - Building Orientation

The Floating Wing Building would face to "SSE" direction so as to win more natural ventilation opportunities in a year as well as to get fresher and cooler air from the sea instead of that from the city.

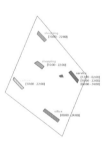

## Wing x Wind

**Step4** - Ventilation Comparison  [Ventilation Modes Diagram and Simulation]

Type A - Tubes Inlet and Outlet
Air vortex and flow backward

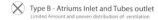

Type B - Atriums Inlet and Tubes outlet
Limited Amount and uneven distribution of ventilation

Type C - Tubes Inlet and Tubes Outlet
Ventilation is well distributed and easily controlled.

**Step5** - Section Conclusion   [Sectional Air Circulation Diagram]

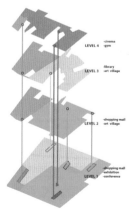

公共空间 PUBLIC SPACE

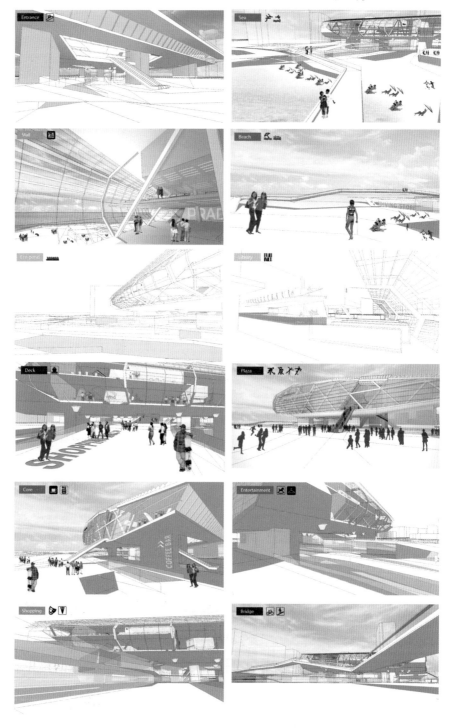

# 风之翼 / WING OF WIND

建构过程 CONSTRUCTION PROCESS

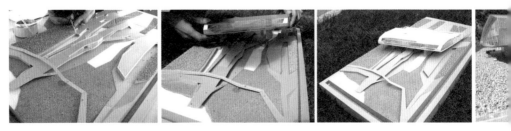

环境透视 PERSPECTIVE

2007 年

## 马来西亚马六甲 / 红树林湿地生态旅游中心
## ECO-TOURIST CENTRE ON THE WETLAND OF MANGROVE MALACCA, MALAYSIA

指导教师 栗德祥 程晓青 邹欢 / 学生 张烨 李煦 国萃 黄逾轩 东京恩 黄怀海 刘芳 睢蔚 刘晓霖 全成浩
Teacher: Li Dexiang, Cheng Xiaoqing, Zou Huan
Student : Zhang Ye, Li Xu, Guo Cui, Huang Yuxuan, Cha Kyung Eun, Huang Huaihai, Liu Fang, Ju Wei, Liu XiaoLin, Jeon SeongHo

2007年"国际建筑工坊"转战马来西亚,在马六甲一个被城市遗忘的角落设计旅游中心。这是我们第一次接触热带地区的住宅设计。马来西亚传统建筑形式独具一格,为学生的设计提供了灵感。地段中还保留有一些自然因素:树木、河流、传统住宅,也成为学生们设计的出发点。从城市空间出发,支离破碎的城市肌理需要缝补,被遗忘的角落需要获得活力进而融入城市,自然因素的保留也为建筑的生态策略提供了可能,从传统建构方式中得到的启发为新技术的应用指出了方向。城市、建筑,新与旧的关系在2007年的大奖方案中得到了全方位的阐释:历史与文化、城市与生活、生态与技术。建筑是技术的结果,更是文化的要求。

In 2007, the AIAC came to Malacca, Malaysia. Students were required to design a tourist center building in a lost corner in Malacca. This was the first time that we embarked on the design of residential building in a tropical area. The unique traditional building of Malaysia inspired the students' designs. There were some natural factors preserved on the site: wood, river and traditional residences, which also served as original idea of the students' designs. Starting from the urban space, they needed to weave the fragmented urban texture, so as to vitalize the abondened corner and integrate it into the city. The preservation of natural factors also provided a possibility for the ecological strategy of the building. The enlightenment from traditional construction guided the direction for the application of new technologies. This winning project in 2007 illustrated completely the relationship between the new and old parts of city and buildings: History and culture, city and life, ecology and technology. Building is a result of technology and more a requirement of culture.

参与院校
清华大学(中国)
巴黎拉维莱特国立高等建筑学院(法国)
米兰理工大学(意大利)
汉阳大学(韩国)
巴勒莫大学(意大利)
马来西亚理工大学(马来西亚)

获一等奖院校:清华大学
清华获奖情况:张烨、李煦荣获一等奖
国萃、黄逾轩荣获二等奖

# 绿色半岛：城市绿点 / GREEN PENINSULA-GREEN URBAN SF

指导教师 栗德祥 程晓青 邹欢 / 学生 张烨 李煦 / 2007 年
Teacher: Li Dexiang, Cheng Xiaoqing, Zou Huan / Student: Zhang Ye, Li Xu / 2007

**教师评语**

该方案以简洁有效的方式把基本居住单元体组合成空间丰富、功能明确、交通便捷的建筑群，同时充分考虑了热带地区建筑的形式和功能特点，使城市中被遗忘的半岛焕发新生。

**COMMENTS**

This project combines the basic residential units into a building complex with colorful spaces, definitive functions and convenient circulation on a simple and effective way. Meanwhile, the project gives full consideration to the architectural form and functional features in the tropical area, and regenerates the forgotten peninsula in the city.

## 概念解读 CONCEPT EXPLANATION

### Green Peninsula-Green Urban Spot
-Comprehensive Development based on urban ecological security network

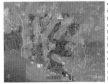

**Terminals**
View from the region aspect, the site, which is demonstrated by a red spot near the sea coast in the left map, is terminals of several green fingers of the whole ecological security system of the city of Malacca. Thus we believe that more green is a must to link the system to the sea.

**Intersection**
Located in the lower reaches of the Malacca river, the green peninsula is not only an extension of the river eco-system, but also plays the role of the intersection of different eco-systems, so as to establish and reinforce the whole ecological security network.

**Oversewing**
Besides the three spots, the small g such as street pa community vegeta should distribute a the city, as well as similar distance to spot, so as to cons grated ecological r Nevertheless, the Malacca city shows

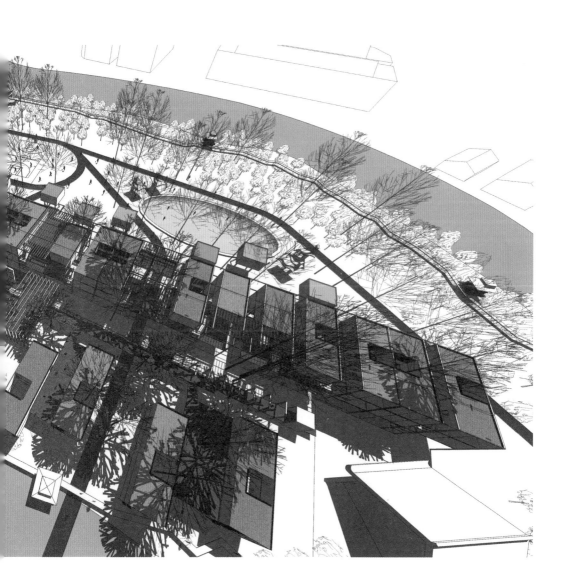

ecological security network is lack of some green spot. The red lines in the map represent the long distance and lower effect between different green spots while the green lines refer to the contrary situation. Therefore, the green peninsula is a must for the city to over-sew its weakened ecological security network.

**Covering**
Defined as an urban green spot, the peninsula reqires more trees and plants to be placed to ameliorate the weak situation at present. Considering the heavy pollution of Malacca River, we decide to extend the area of mangroves, as well as planting new trees, so as to optimize the whole environment.

**Get-together**
The green peninsula not only plays the role of the green urban spot, but also provides the inhabitants of the neighborhood with a public space or green city park for leisure and communication. People can enjoy the fresh air, beautiful environment and excited life here, as well as join the community activities.

# 绿色半岛: 城市绿点 / GREEN PENINSULA-GREEN URBAN SP

## 概念生成 CONCEPT GENERATION

**Filtrate**
We extremely careful with respect to the important and valuable existing elements, such as the big trees, cinema and Kampung Morten.

**Connect**
We especially focus on Kampung Morten, and try to consider about the two parts at the same time. What is the relationship in-between?

**Divide**
We hope that the buildings will create the pattern of the site from nature to public space and then to private space.

**Interpose**
We attempt to establish more connections between the two sides, according to different types of people, which will make the site to be an accessible neighborhood.

**Origine**
In order to set aside more green land, we try to make the houses as concentrated as possible. The Baba and Nyonya house is a good antetype.

**Keep**
Old things with memory and innate energy have great potential to define local identity. The silhouette of the old cinema possesses such functions. An initial intervention is needed to trigger the series of changes.

**Enlarge**
Defined as a "green spot" in the city of Malacca, the mangroves around need to be enlarged, contributing to making the peninsula a real green spot.

**Interact**
Holding the notion that "My house is my neighborhood, my neighborhood is my house", we don't view our project as building but as social environment, where different people can actually go through the spaces, where they can actually meet one another.

**path system**
We strive to classify by groups the future residents of the site. We also classify each type of individuals according to the services within the overall concept that he or she will use most habitually. The five entrances of the site combine different types of people, resulting in various defined paths to establish a integrated path system.

community activity     jogging

**traffic analysis**

 +  +  +  +

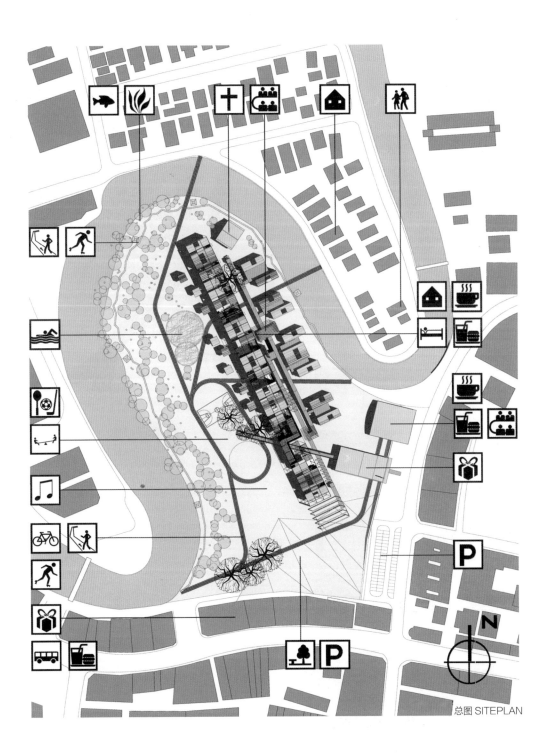

总图 SITEPLAN

# 绿色半岛：城市绿点 / GREEN PENINSULA-GREEN URBAN SP

公共空间 PUBLIC SPACE

program pattern

2007年/马来西亚马六甲/红树林湿地生态旅游中心 | 117

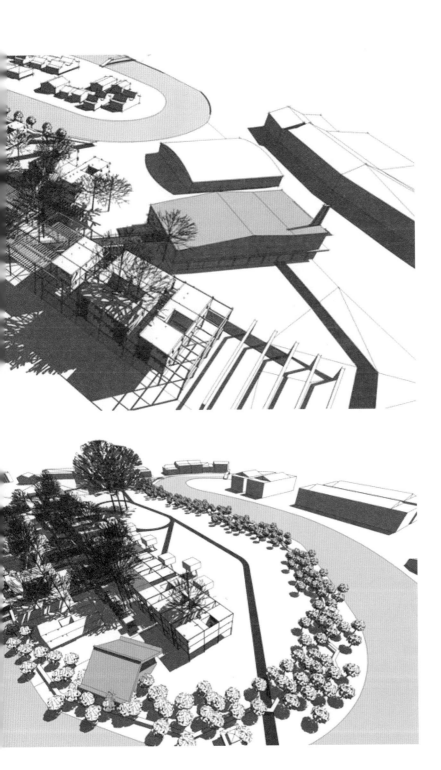

# 绿色半岛：城市绿点 / GREEN PENINSULA-GREEN URBAN SP

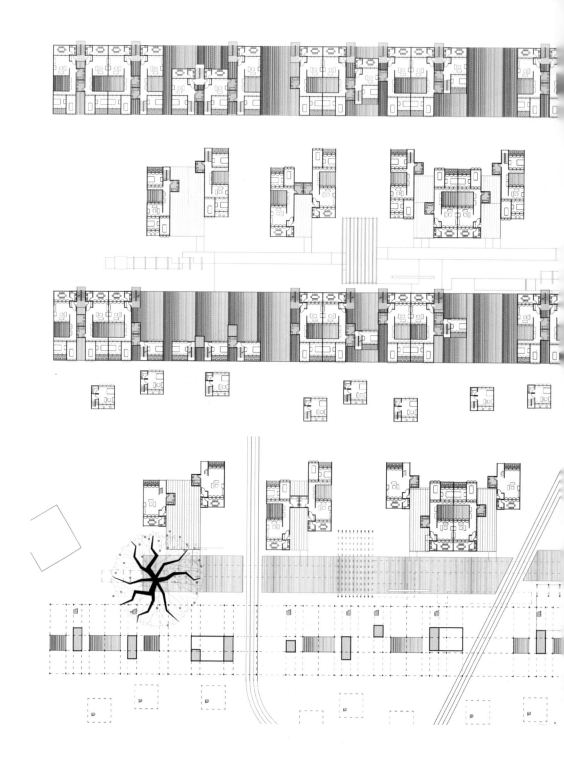

**third floor plan**

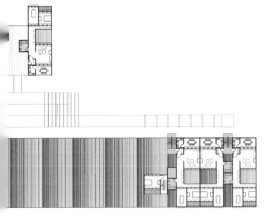

**second floor plan**

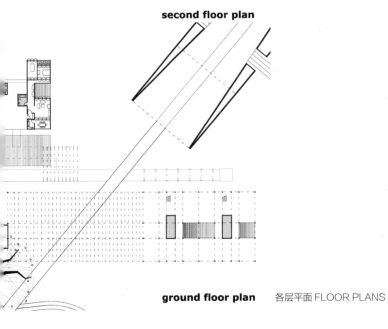

**ground floor plan**　各层平面 FLOOR PLANS

# 绿色半岛：城市绿点 / GREEN PENINSULA-GREEN URBAN SPACE

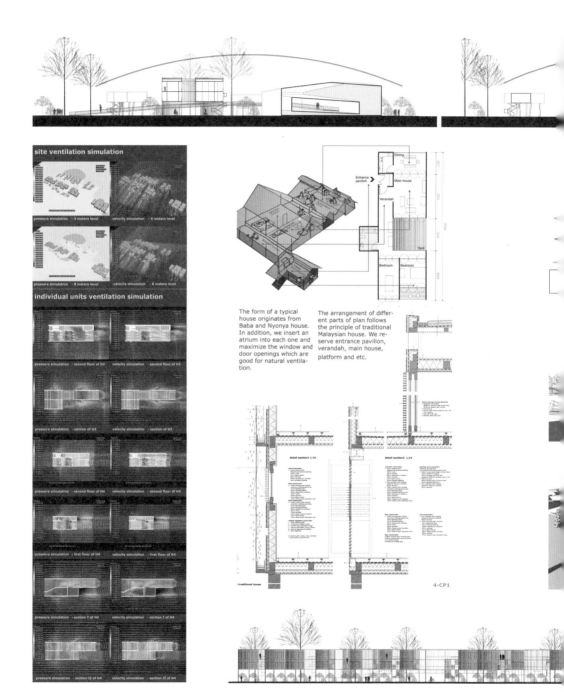

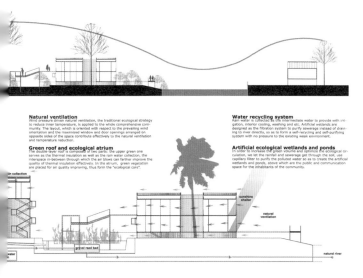

**Natural ventilation**
Wind pressure driven natural ventilation, the traditional ecological strategy to reduce inner temperature, is applied to the whole comprehensive community. The layout, which is oriented with respect to the prevailing wind orientation and the maximized window and door openings arranged on opposite sides of the space contribute effectively to the natural ventilation and temperature reduction.

**Green roof and ecological atrium**
The double-layer roof is composed of two parts: the upper green one serves as the thermal insulation as well as the rain water collection, the interspace in-between through which the air blows can farther improve the quality of thermal insulation effectively. In the atrium, green vegetation are placed for air quality improving, thus form the "ecological core".

**Water recycling system**
Rain water is collected as the intermediate water to provide with irrigation, interior cooling, washing and etc. Artificial wetlands are designed as the filtration system to purify sewerage instead of draining to river directly, so as to form a self-recycling and self-purifying system with no pressure to the existing weak environment.

**Artificial ecological wetlands and ponds**
In order to increase the green volume and optimize the ecological circulation, we let the rainfall and sewerage get through the soil, use capillary filter to purify the polluted water so as to create the artificial wetlands and ponds, above which are the public and communication space for the inhabitants of the community.

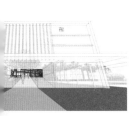

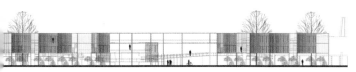

west elevation

技术设计 TECHNICAL DESIGN

# 林中生活 / LIVING IN FOREST

指导教师 栗德祥 程晓青 邹欢 / 学生 国萃 黄逾轩 / 2007 年
Teacher: Li Dexiang, Cheng Xiaoqing, Zou Huan / Student: Guo Cui, Huang Yuxuan / 2007

**教师评语**
想法浪漫而具有诗意——像鸟一样栖息在树林中，利用地段上现有的茂密植被，营造出一个架空的生活社区。为避免单元式的住宅平面单调乏味，从解决日照和遮阳的建造技术上去寻求空间变化。

**COMMENTS**
The project has a romantic and poetic idea - as if a bird lives in the forest. It creates an overhead living community over the strength of dense vegetation in the site. To avoid the monotonous of unit-typed residential plan, the project seeks for spatial variation by sunlight and sunshade building techniques.

概念解读 CONCEPT EXPLANATION

2007年/马来西亚马六甲/红树林湿地生态旅游中心 | 123

# 林中生活 / LIVING IN FOREST

二层平面 2F PLAN

2007年/马来西亚马六甲/红树林湿地生态旅游中心

一层平面 1F PLAN

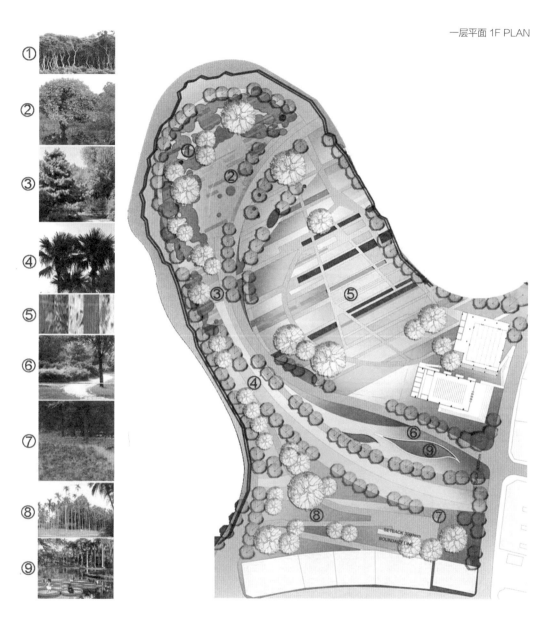

立面 FACADE

# 林中生活 / LIVING IN FOREST

## 形体生成 TRANSFORMATION

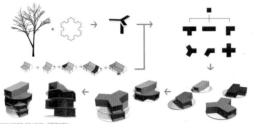

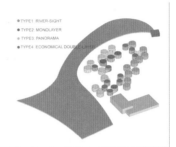

PARTICULAR DESIGN OF TYPE

As a typical and original unit, R sight can represent all other type houses. Therefore, it is chosen model to carry on detail ideas o design. Sections are used to study space and aeration, and 3D-me are built to study its space use. is more, with the help of professi software, the real aeration of h is imitated based on the weather of the site. Other types of house similar to the typical type in ecolo design, and will not be discusse particular as type1.

CONCEPT OF UNIT : GROWTH
The two main factors of concept are trees, the texture of the nature, and local houses, the texture of the culture. There is a common point between the two, and that is GROWTH. Trees grow to pursue the sunshine, while local houses of Malaysia grow to make more spaces. In this historical byland full of trees, it is better to obey the law of nature and culture to creat a new type house. The result will fit the site in a fresh way, because it is different but still comes from characters of the site.

## 各层平面 FLOOR PLANS         环境模拟 SIMULAT

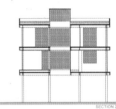

SECTION 1(1:150)        SECTION 2

●TYPE1: RIVER-SIGHT(1:150)
Houses of this type are mainly located besides the river, where there is a nice river scape. Therefore, the design is open to the river and with large size. Moreover, verandas of different fuction are disposed alongwith the indoor space, such as porch, guest veranda, family varanda and daily varanda, which all come from custom of Malaysian life style. Revolvable doors can both separate and connect the indoor and outdoor spaces.

ANALYSIS OF AERATION
VELOCITY OF WIND          PRESSURE O

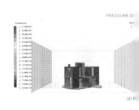

●TYPE2: MONOLAYER(1:150)
Houses of this type are mainly located in the central area of site, where the sight is less satisfied. Therefore, the design is with small size and only monolayer for each family. Verandas are disposed as well but fewer. Horizontal aeration is acheved by open windows, while vertical aeration is taken place in the patio. Alliance of the two methods increase the efficiency of indoor aeration.

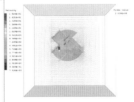
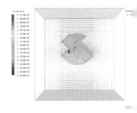

●TYPE3: PANORAMA(1:150)
Houses of this type are mainly located in the mangrove, which surround the houses. All sides of the house have nice but different sight. In order to take advantage of this factor, the house is designed in the concept of panorama which means residence can enjoy scene of 360 degrees. Veranda plays important part in the touch with surronded nature like other types, and the number of them is large enough to satisify the needs of viewing.

●TYPE4: ECONOMICAL DOUBLE-LAYER(1:150)
Houses of this type are mainly located in the central area of site, where the sight is less satisfied. The design is economical in size like type2, but it has better condition than type2 because there are more verandas and flexible outdoor spaces.

2007年/马来西亚马六甲/红树林湿地生态旅游中心 | 127

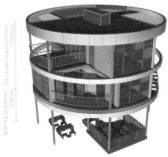

VIEWS OF THE HOUSE

# 林中生活 / LIVING IN FOREST

## 技术设计 TECHNICAL DESIGN

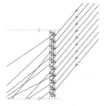

**REGULABLE SYSTEM**

Healthy light environment is important for a green house, And sun protectuion solutions have a role to optimize the building environment and save the energy. This programe use the local material banboo to make the prinoting shutter.One side it changes the sun ray into diffuse reflection. The other side, it economizes the price.

**A  LIANE SHUTTER (NODE 1)**
Shutter can prevent houses from strong sunshine. Liane,which grows on the shutter, can improve the fuction and bring about green to the houses.

**B  SOLAR THERMAL UTILIZATION**
Malayisa has abundant solar resource.The energy-technology concept specified solar thermal utilization which feed into an interim storage system for non-potable water and heating.

**C  CONTROLLABLE SHTTER (NODE 1)**
Controllable shutter can be used to adjust the use of the wind tower. When the shutters are open, airflow of the well can improve the aeration of the house. When they are closed, vertical airflow stops.

**D  BLACK SURFACE OF THE WIND TOWER**
Black surface can absorb more heat from sunshine than the undertone one. The hotter the top is, the more efficient the wind tower will be.

**E  RAIN GATHERING (NODE 1)**
Rainwater is gathered systematically in the roof. Some of the rainwater is used to supply the water to the green roof, and the other flow in to the bos gutter.

**F  GREEN ROOF (NODE 3)**
Planting some grass on the roof could insolute the heat which comes from the solar and use the rainwater resonably. On the other side it represents the conception more visibly.

**G  PHOTOVOTAIC (NODE 1)**
The potovotaic system produces more electricity than is required for this energy-efficient house.

**H  BANBOO PIVOTING SHUTTER (NODE 1)**
The banboo shutter can pivot smoothly. When they are rotated to 90, the inter space changes into outdoor space. Then the natural airflow supplies the comfortable temperature.

**I  NATURAL AIRFLOW (NODE 1)**
Taking advantage of the natural wind, the house uses a lot of shutters to free the airflow. Natural airflow can cool down the temperature and take away moisture.

**J  WIND TOWER**
High wind tower in the house can improve the efficiency of aeration, especially in the days with little wind.

**K  MOISTURE ABSOPTION WALL**
Walls indoor use meterial that can absorb moisture, so that the humidity of the air can become smaller. It is good to the health of residents.

**L  WATER COOLING WALL (NODE 2)**
Cool water pumped from the underground is delievered up to the house and cool down the indoor temperature. Cooling water wall is located in west side of the house, where there are few windows, and can cover most part of the house.

**M  DOUBLE FLOOR  VENTILATION (NODE 2)**
Cool air from the underground can be delievered to the house through the pipe well. And the indoor space uses double floor  to transport the air to cool down the temperture.

**N  LIANE WALL (NODE 1)**
The veranda is surrounded by liane, which can shade the veranda from the sunshine. Moreover, the wind blowing into the house can be cooled down by the green plant. It is also a scenery to the house.

**O  LIANE SUBSTRAT (NODE 1)**
On the circle girder, the substrat of liane is placed.

**P  BANBOO SLIDING SHUTTER ()**

**Q  PIPELINE WELL**
The well coming down from the upper house has a lot of pipeline in it, such as water, electricity, air and so on. It is covered by green plants on the ground level, which is also a scenery of landscape.

**ECOLOGY OF COMMUNITY**

**SECTION INDICATING AIRFLOW**

Environmentally-friendly community is a characteristic of this project. Greening, natural ventilation, cool wall, heat storage system and photovoltaic generation and so on are all rhe effective methods. Natural airflow is first considered as an important element in particular. multipurpose underground interlayer, approximately 90 sq.m, is constructed running through the house. This interlayer works as a cool tube by the agency of the stable underground temperature 15-20 C at all year around. The fresh air handing achine additionally arranges the fresh air and send ir out to the under floor of each room and the cool wall in the house. And then that air is slowly blown out from the window side outlet, naturally drawn upward through the vertical void named "Wind Tower" to the exhaust outlet atop the house.  ealthy light environment is important for a green house, And sun protectuion solutions have a role to optimize the building environment and save the energy.

**ECO POND**

EnMalaysia is a tropical country, and there is abundant water resource around the site. So water is on of the most important elements in this project, eco pond is planed in the site. Different kinds of hydrophytes are planted to purify the rainwater. And pond effectively serves some other multiple functions:
Keeping the weather out;
Providing necessary shade whilst preserving cross ventilation;
Connecting all the various spaces and creating an interface to corporate landscaping;
Forming a high quality gathering space for contemplation and decision making;
Making the air cooler across the water evaporation;

**EFFECT OF REDUC( HOUSE)**

Many trpical vegeta micro-climatic much difference of appro to sunlight. It is also building. The micro-e Calculation assume A 75% reduction of And the next generi reduce.

## 公共空间 PUBLIC SPACE

GREEN ROOF

BABBOO SHUTTER

WATER COOLING WALL

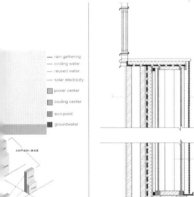

- rain gathering
- cooling water
- reused water
- solar electricity
- power center
- cooling center
- eco-pond
- groundwater

...ON AND CO2 EMISSION (PER

by step. They could make the
...de effect provides temperature
...led areas and areas exposed
...re passed through the site and
...sults for lifestyles.

...plantings (from survey results),
...oning and co2 emissions could

NODE 2     WATER COOL

2006 年
## 法国马赛 / 军事遗迹改造与再利用
## RENOVATION AND REUSE OF MILITARY HERITAGE, MARSEILLE, FRANCE

指导教师 栗德祥 程晓青 邹欢 / 学生 兰俊 李德熙 王静 陈洁 丁依菲 高珊 葛鑫 黄俊杰
Teacher: Li Dexiang, Cheng Xiaoqing, Zou Huan
Student: Lan Jun, Lee Duck Hee, Wang Jing, Chen Jie, Ding Yifei, Gao Shan, Ge Xin, Huang Junjie

2006年/法国马赛/军事遗迹改造与再利用 | 131

中国学生听到马赛时的第一反应就是柯布西耶、马赛公寓。带着这个挥之不去的现代建筑情节，2006年的3月我们来到了法国南部重要城市马赛，然而我们的任务比柯布西耶更为复杂：在马赛这个历史悠久的城市对一个老旧的兵营街区进行改造。兵营位于一个沉闷的街区，周边建筑乏善可陈。如何为一个平庸的街区内注入新的活力？如何与雄壮严肃的历史建筑相为邻？如何用一个为年轻人提供居住的学生公寓来活跃城市生活？城市公共空间，交流、渗透，建筑组群的姿态，成为这个题目的研究重点。地段内有一座现存建筑，没有太多的历史价值，有趣的是中国学生多数选择了保留，而欧洲学生大多选择了拆除，文化差异可见一斑。

When Marseilles was mentioned, the first reaction of Chinese students was Unite d'Habitation à Marseille by Le Corbusier. With such a lingering affection of modern architecture, we came to Marseilles, an important city in the south of France in March 2006. However, our task was more complicated than Le Corbusier: To transform an old barrack block in such a time-honored city like Marseilles. The barrack was located at a tedious street, without any nice or unusual buildings in the surroundings. How to inject new life into such a mediocre block? How to make it a neighbor of the majestic historical buildings? How to enliven the city life through a student apartment building? The research for this project was urban public space, exchange, penetration and forms of building complex. There was an existing building in the site, but it didn't have too much historical value. What was interesting was that most Chinese students preserved it while a majority of Europan students chose to demolish it. The cultural difference was quite evident.

参与院校
清华大学（中国）
巴黎拉维莱特国立高等建筑学院（法国）
米兰理工大学（意大利）
汉阳大学（韩国）
那不勒斯大学（意大利）
巴勒莫大学（意大利）
达姆施达德大学（德国）

获一等奖院校：巴黎拉维莱特国立高等建筑学院
清华获奖情况：兰俊荣获二等奖
李德熙荣获三等奖
王静荣获优秀奖

# 音乐节点连廊 / CORRIDOR OF MUSIC

指导教师 栗德祥 程晓青 邹欢 / 学生 兰俊 / 2006 年
Teacher: Li Dexiang, Cheng Xiaoqing, Zou Huan / Student: Lan Jun / 2006

**教师评语**
该方案大胆利用老建筑形成内院式的公共广场，新旧建筑有机对话，空间整齐，具有韵律感。

**COMMENTS**
This project boldly utilizes the old buildings to form a courtyard as public square. The new and old buildings have an organic dialogue, creating a neat space and a sense of rhythm.

2006年/法国马赛/军事遗迹改造与再利用 | 133

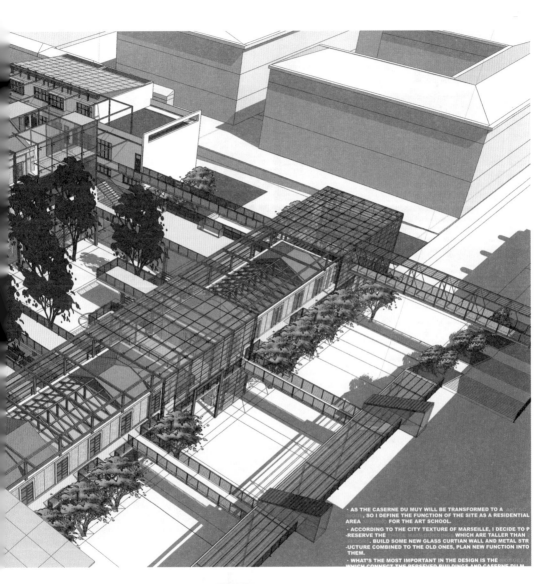

- AS THE CASERNE DU MUY WILL BE TRANSFORMED TO A _____, SO I DEFINE THE FUNCTION OF THE SITE AS A RESIDENTIAL AREA _____ FOR THE ART SCHOOL.
- ACCORDING TO THE CITY TEXTURE OF MARSEILLE, I DECIDE TO P -RESERVE THE _____ WHICH ARE TALLER THAN _____. BUILD SOME NEW GLASS CURTIAN WALL AND METAL STR -UCTURE COMBINED TO THE OLD ONES, PLAN NEW FUNCTION INTO THEM.
- WHAT'S THE MOST IMPORTANT IN THE DESIGN IS THE _____ WHICH CONNECT THE RESERVED BUILDINGS AND CASERFE DU M

环境分析
ENVIRONMENT
ANALYSIS

# 音乐节点连廊 / CORRIDOR OF MUSIC

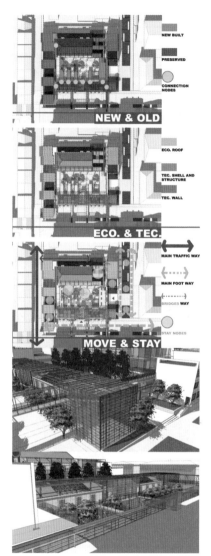

概念分析 CONCEPT ANALYSIS

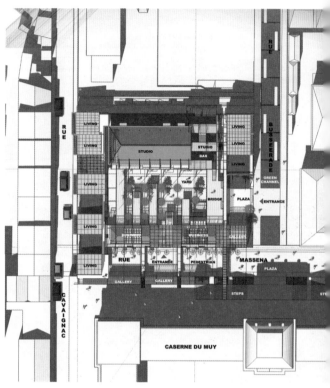

总平面 SITE PLAN

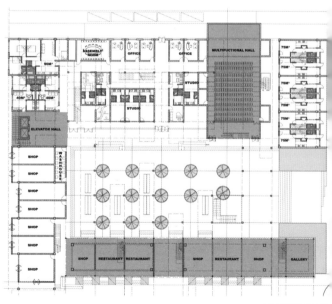

一层平面图 1F PLAN

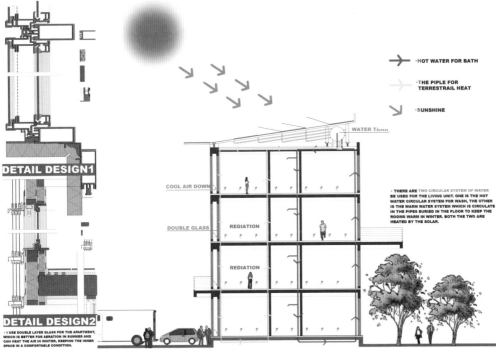

节点设计 DETAIL DESIGN

剖面设计 SECTION DESIGN

# 音乐节点连廊 / CORRIDOR OF MUSIC

### 技术设计
### TECHNICAL DESIGN

· USE A WELL DESIGNED GLASS CURTAIN WALL TO REFORM THE SHAPE OF THE OLD BUILDINGS NEXT TO THE STREET. THE CURTAIN WELL CAN SHIELD THE HARM LIGHT AND RESTORE THE HEAT WHICH CAN BE USED FOR THE OLD BUILDIGS AS WELL.

· AS MENTIONED ABOVE THE SPECIAL OF THIS DESIGN IS A LOT OF BRIDGES BETWEEN THE CASERNE DU MUY AND OUR SITE, THE SHAPE OF WHICH IS LIKE THE MUSIC NOTE ON THE SCREEN. AS THE PLATFORMS OF CASERNE DU MUY AND REBUILT BUILDINGS ARE ON DIFFERENT HEIGHT LEVEL, THE BRIDGE NEED A SLOPE. I PUT THE SLOPE IN THE GLASS CURTIAN WALL BOX, WHICH TOGETHER WITH THE PLATFORM BELOW, ENTICH THE SPACE.

· I ADD OUT-BALCONY TO THE BUILDINGS NEXT TO THE STREET TOGETHER WITH THE CURTIAN WALL, WHICH MAKE THE OLD BUILDING DYNAMIC AND SUPPLY THE PASSIBILITY OF COMMUNICATION BETWEEN THE BUILDINGS AT ONE SIDE AND THE OTHER, ALSO THE PEOPLE IN THE FORTHCOMING ART SCHOOL AND THE COMMUNITY.

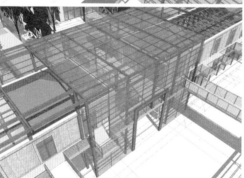

### 剖面 SECTION

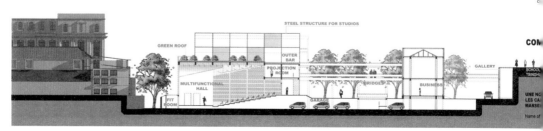

2006年/法国马赛/军事遗迹改造与再利用 | 137

# 呼吸自然 /BREATHING WITH NATURE

指导教师 栗德祥 程晓青 邹欢 / 学生 李德熙 / 2006 年
Teacher: Li Dexiang, Cheng Xiaoqing, Zou Huan / Student: Lee Duck Hee / 2006

**教师评语**
将生态技术措施合理有效地运用到建筑当中，绿化景观与雨水净化结合，垂直绿化与建筑遮阳结合，让公共空间成为呼吸自然的场所。

**COMMENTS**
The ecological techniques are applied in the building reasonably and effectively. The combination between vegetation landscape and rain purification, vertical greening and building sunshade, makes the public space a place of breathing in the nature.

概念解读 CONCEPT EXPLANATION

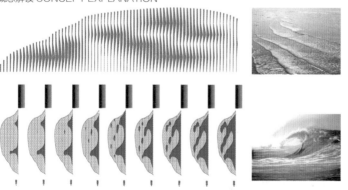

Light Box

2006年/法国马赛/军事遗迹改造与再利用 | 139

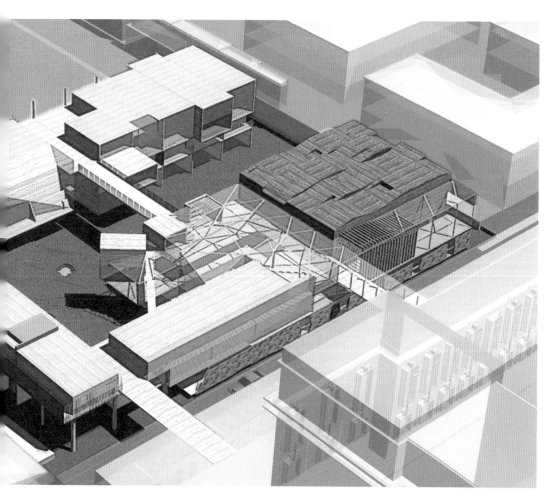

# 呼吸自然 / BREATHING WITH NATURE

技术设计 TECHNICAL DESIGN

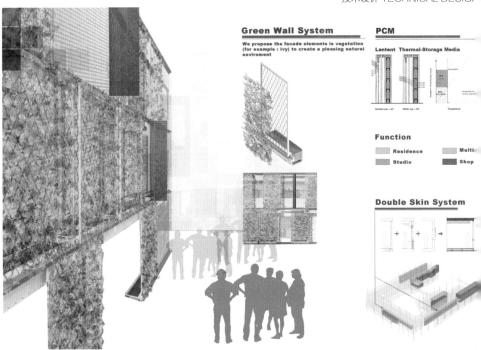

平剖面设计 PLAN AND SECTION DESIGN

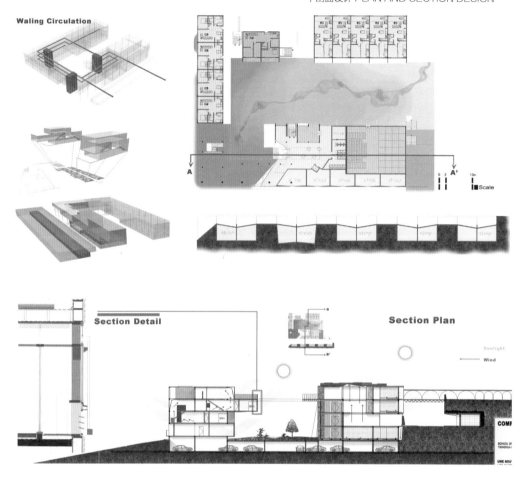

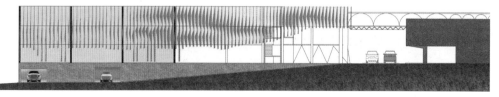

# 活力社区 / ACTIVE COMMUNITY

指导教师 栗德祥 程晓青 邹欢 / 学生 王静 / 2006 年
Teacher: Li Dexiang, Cheng Xiaoqing, Zou Huan / Student: Wang Jing / 2006

### 教师评语
该方案保留老建筑的基础，巧妙利用地下空间，营造废墟考古式的空间氛围，为公共广场带来亮色。建筑处理手法细腻，构造研究深入。

### COMMENTS
This project reserves the foundation of old buildings and takes advantage of the underground space cleverly to create an atmosphere of archaeological ruins, bringing a highlight for public square. The approach of architectural treatment is exquisite and the detail study is profound.

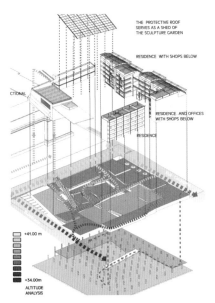

概念解读 CONCEPT EXPLANATION

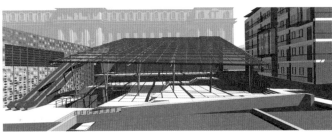

2006年/法国马赛/军事遗迹改造与再利用 | 143

城市肌理 THE URBAN TEXTURE　　城市开放空间 THE URBAN OPEN

# 活力社区 / ACTIVE COMMUNITY

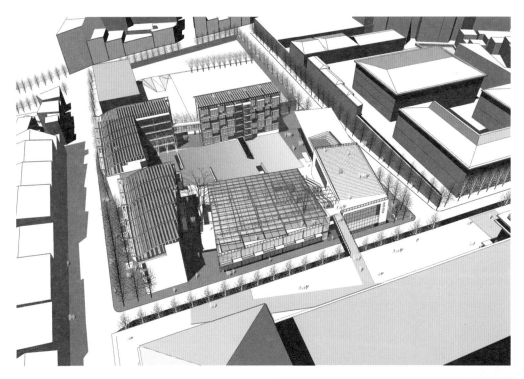

被动技术设计
PASSIVE DESIGN TECHNIQUES

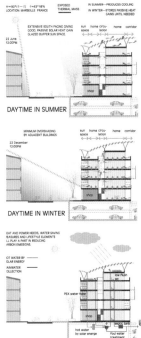

2006年/法国马赛/军事遗迹改造与再利用 | 145

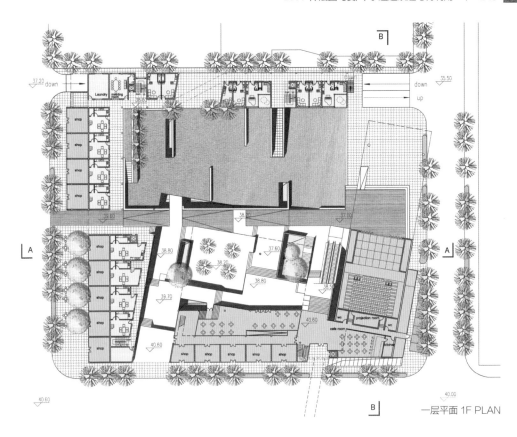

一层平面 1F PLAN

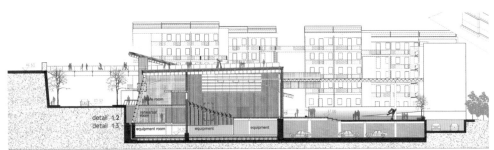

B-B 剖面 SECTION

A-A 剖面 SECTION

2005 年

# 意大利巴勒莫 / 旧港区更新计划
# RENEWAL OF THE OLD PORT PALERMO, ITALY

指导教师 栗德祥 程晓青 邹欢 / 学生 霍振舟 安德雷 梁思思 刘聪 许阳 范小棣 张阳 张婷
Teacher: Li Dexiang, Cheng Xiaoqing, Zou Huan
Student : Huo Zhenzhou, Andres Rocha Pardo, Liang Sisi, Liu Cong, Xu Yang, Fan Xiaodi, Zhang Yang, Zhang Ting

多种文化的交织融合形成了意大利西西里复杂深厚的城市历史，地中海的阳光照射在深宅老巷，建筑系的学生迷失在扑簌迷离的光影中……2005年的设计地段位于西西里城市中心，昔日的渔港，这里是西西里生活开始的地方，承载着西西里厚重的历史。随着城市的发展，渔港已经失去了原有的功能，鱼市也已废弃，功能主义盛行时期建设的道路破坏了城市肌理，亟需改造，缝合碎裂的城市空间，创造舒适宜人的城市环境，重现港口昔日风采。学生的方案从整合城市空间入手，运用古典的手法和现代的语汇创造新的建筑组群，同时对紧邻的古堡遗址表达了充分的敬意。

The combination of various cultures constituted the complicated and rich urban history of Sicily, Italy. When the sunlight of the Mediterranean blazed down on those mysterious houses and old lanes, students of architecture got lost in the blurred light and shadow... In 2005, the site was located in the center of Sicily, the fishing harbor at the ancient times. It represented the origin of Sicilian life and long history of Sicily. The fishing harbor was deprived of its original functions along with the urban development, so was the fish market. The urban texture was destroyed by the roads built in the prevalence of functionalism. Transformation was urgently needed to make up the broken urban spaces, create a comfortable and pleasant urban environment, and reappear the disappeared grandeur of the port. Starting with the integration of urban space, the students' projects created new complexes through classic techniques and modern architectural vocabularies, and also expressed their respect to the castle ruins nearby.

参与院校
清华大学（中国）
巴黎拉维莱特国立高等建筑学院（法国）
米兰理工大学（意大利）
汉阳大学（韩国）
那不勒斯大学（意大利）
巴勒莫大学（意大利）

获一等奖院校：米兰理工大学
清华获奖情况：霍振舟荣获优秀奖

# 滨海广场 /PLAZA BY THE SEA

指导教师 栗德祥 程晓青 邹欢 / 学生 霍振舟 / 2005 年
Teacher: Li Dexiang, Cheng Xiaoqing, Zou Huan / Student: Huo Zhenzhou / 2005

**教师评语**
该方案用古典欧洲城市广场的设计手法，在巴勒莫这样一个历史悠久、文化丰厚的城市中心延续其空间格局。

**COMMENTS**
Using the forms of classic European city square, this project continues the space pattern in the urban center in Palermo with long history and rich cultures.

环境分析 ENVIRONMENT ANALYSIS

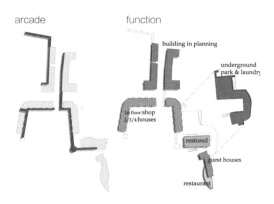

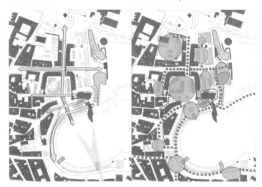

# 滨海广场 /PLAZA BY THE SEA

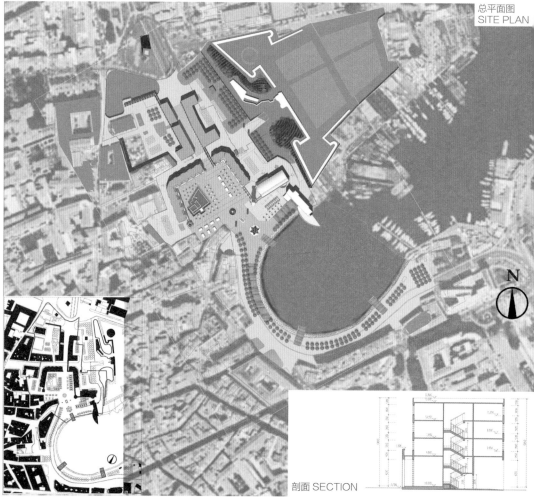

总平面图
SITE PLAN

剖面 SECTION

2005年/意大利巴勒莫/旧港区更新计划 | 151

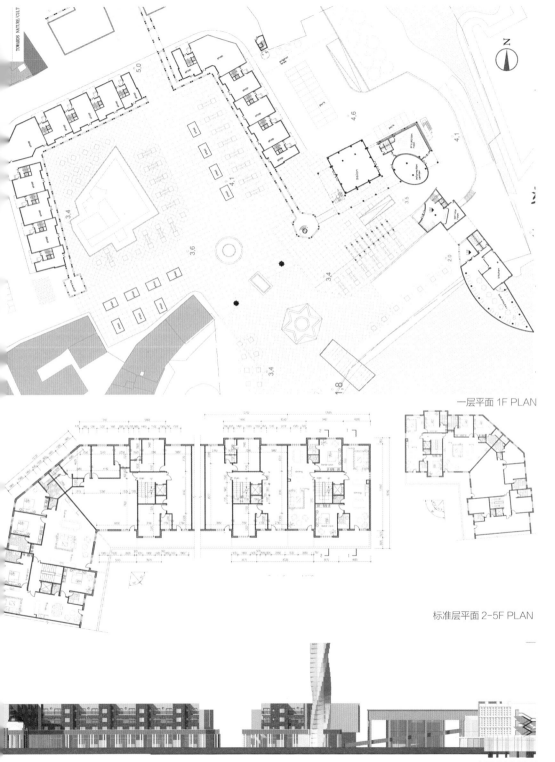

一层平面 1F PLAN

标准层平面 2-5F PLAN

立面 FACADE

2004 年
# 韩国首尔 / 现代艺术博物馆
# MODERN ART MUSEUM
# SEOUL, R.O. KOREA

指导教师 栗德祥 程晓青 邹欢 / 学生 雷亮 申吉秀 郑天 初腾飞 喻晓 高博 李乃桢 余知衡 霍顺利 赖延璨 李海琳娜·毕玛丽 白洁 陈瑾羲 吴懿 郑珊珊 邓健 杨晓昕 腾静茹 郭佳

Teacher: Li Dexiang, Cheng Xiaoqing, Zou Huan
Student : Lei Liang, Shin Gil Soo, Zheng Tian, Chu Tengfei, Yu Xiao, Gao Bo, Li Naizhen, Yu Zhihe
            Huo Shunli, Lai Yentsan, Li Haiyan, Helene Bihlmaier, Bai Jie, Chen Jinxi, Wu Yi,
            Zheng Shanshan, Deng Jian, Yang Xiaoxin, Teng Jingru, Guo Jia

2004年"国际建筑工坊"来到韩国首都首尔。首尔是一个混杂而迷人的城市,设计地段紧靠景福宫——昔日的韩国王宫。与王宫一墙之隔的地段周围,传统的城市肌理依然大量存在,在这些传统街巷中散布着丰富多彩的艺术家工作室和画廊,传统与现代的交织形成了这里的城市空间特色。像大多数的亚洲城市一样,新与旧并存甚至直接对峙在城市空间中形成了各种奇妙的张力,如何处理这些张力就成为建筑设计成败的关键。加强?缓和?折衷?取消?学生们的方案给出了各种回答,对建造技术的关注也是2004年"国际建筑工坊"设计的特点,与东方传统建筑强调建构不谋而合。

AIAC came to Seoul, the capital of R.O. Korea, in 2004. Seoul is a mixed but fascinating city. The design site was close to the Gyeongbokgung Palace - the old Palace of R.O. Korea. The traditional urban texture still remained significant in the surroundings of the site, just beside the Palace. In these traditional streets and alleys, there were colorful artist studios and galleries. The combination of tradition and modernity created the features of urban space. Like most Asian cities, the coexistence and even direct standoff of the old and new shaped all kinds of wonderful tension. Hence, the key of architectural design rested with how to treat these tensions. Strengthen? Alleviate? Compromise? Cancel? Students gave different answers in their projects. The concern for construction technology was also one of the features in the designs of AIAC, which coincided with the oriental traditional buildings' characters.

参与院校  
清华大学(中国)  
巴黎拉维莱特国立高等建筑学院(法国)  
米兰理工大学(意大利)  
汉阳大学(韩国)  
那不勒斯大学(意大利)  

获一等奖院校:汉阳大学  
清华获奖情况:雷亮、申吉秀荣获优秀奖  

# 走向文化与自然 / TOWARDS CULTURE & NATURE

指导教师 栗德祥 程晓青 邹欢 / 学生 雷亮 申吉秀 郑天 / 2004 年
Teacher: Li Dexiang, Cheng Xiaoqing, Zou Huan / Student: Lei Liang, Shin Gilsoo, Zheng Tian / 2004

自然，意指生态，物理环境和人的原始感受；
文化，意指文明，风俗传统和人的精神感觉。
基地周边：汉城市中心，西邻韩国"紫禁城"——景福宫，其余三面为传统民居。
方案重点：自然与文化。

Nature is ecology, physical environment and human's original feeling.

Culture is civilization, tradition and human's spiritual feeling.

The key point of the project is nature and culture.

The site locates in the center of Seoul. Facing west stands Gyeongbokgung Palace - the "Forbidden City" of R.O. Korea, with traditional residential buildings surrounding the site.

**教师评语**
该方案用完全覆盖的绿化屋顶向近在咫尺的韩国故宫——景福宫致敬。用自然去尊重文化与历史也许是最安全的办法。

**COMMENTS**
This project applies the fully-covered greening roof to express the salute to Gyeongbokgung Palace - the Palace of R.O. Korea nearby. Maybe it is the safest way to respect cultures and history by ways inspired form the nature.

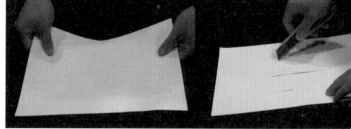

2004年/韩国首尔/现代艺术博物馆 | 155

# 走向文化与自然 / TOWARDS CULTURE & NATURE

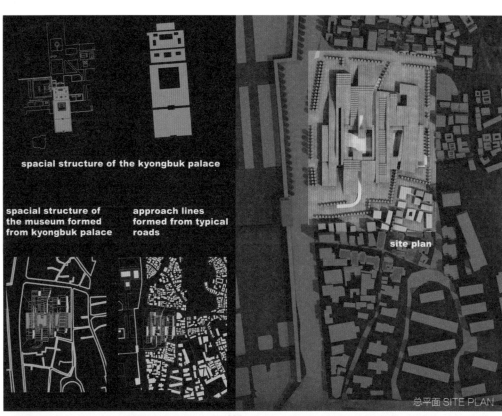

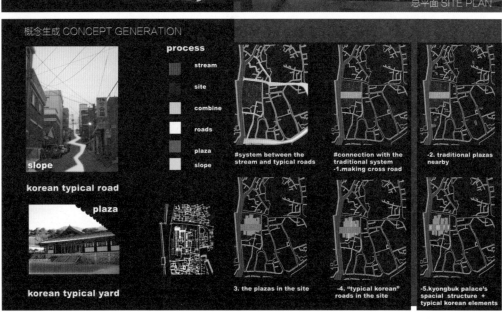

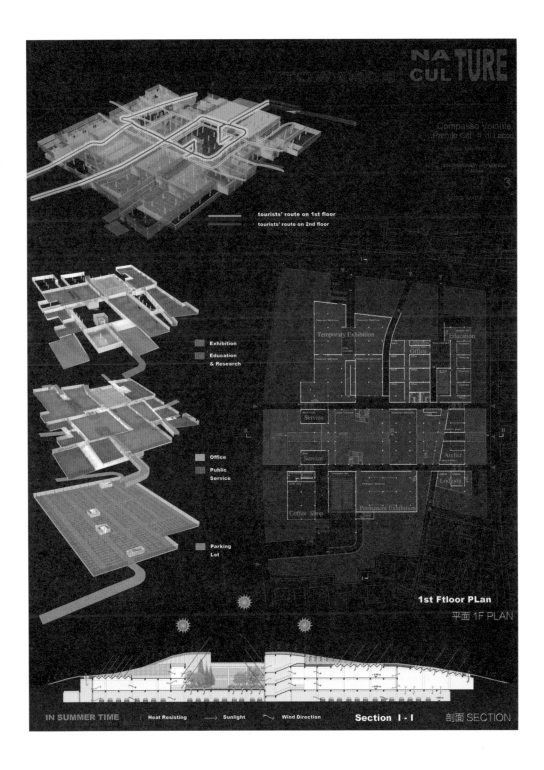

2003 年
# 中国北京 / 高层生态写字楼
# ECO-HIGHRISE, BEIJING, CHINA

指导教师 栗德祥 程晓青 / 学生 庞聪 储以平 李楠 陈晶 张维 田宏 卜骁骏 解霖 赵扬
Teacher: Li Dexiang, Cheng Xiaoqing
Student: Pang Cong, Chu Yiping, Li Nan, Chen Jing, Zhang Wei, Tian Hong, Bu Xiaojun, Xie Lin, Zhao Yang

2003年/中国北京/ 高层生态写字楼 | 159

高层，高层，雄伟的高层！中国快速发展的城市建设，北京打造国际大都市的雄心，为"国际建筑工坊"提供了实现梦想的机会。2003年的地段选在北京四环边上，任务是高层办公楼。这是"国际建筑工坊"第一次也是唯一一次做高层项目。全力以赴迎接奥运的北京看上去是那么地激动人心，各国学生们都在畅想未来的建筑、空间、技术。可持续发展是2003年的热门概念，所有的方案中都涉及生态技术，虽然只是畅想，但是年轻人的热情理想难能可贵。

High-rise, high-rise, majestic high-rise! The rapid development of urban construction in China and Beijing's ambition of being a cosmopolis fostered an opportunity of approaching its dream for the AIAC Design Studio. In 2003, the selected site was located near the West 4th Ring Road, Beijing. The project was to design a high-rise office building. This was the first and the only one time that AIAC embarked on a high-rise building project. Beijing, which were going all out to welcome the Olympic Games, looked quite exciting. Students of all countries were weaving their fantasy for future architecture, space and technique. Sustainable development was a popular concept in 2003, so all projects involved ecological technologies. Although it was a fantasy, the passionate ideas of these young people were commendable.

参与院校
清华大学（中国）
巴黎拉维莱特国立高等建筑学院（法国）
米兰理工大学（意大利）
汉阳大学（韩国）
那不勒斯大学（意大利）

获一等奖院校：米兰理工大学
清华获奖情况：庞聪、李楠、储以平荣获优秀奖

# E 空间 / E-SPACE

指导教师 栗德祥 程晓青 / 学生 庞聪、褚以平、李楠 / 2003 年
Teacher: Li Dexiang, Cheng Xiaoqing / Student: Pan Cong, Chu Yiping, Li Nan / 2003

**教师评语**
该方案在地段苛刻的条件下，通过日照和落影计算来形成建筑体型，以期达到建筑各方面影响作用的最优化。

**COMMENTS**
This project forms a building geometry by the calculation of sunlight and shadow under the demanding conditions of the site, in hope of optimizing all influential effects of the building.

总平面 SITE PLAN

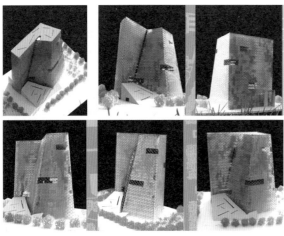

2003年/中国北京/ 高层生态写字楼

# E 空间 / E-SPACE

ATRIUM PERSPECTION

中庭透视 ATRI

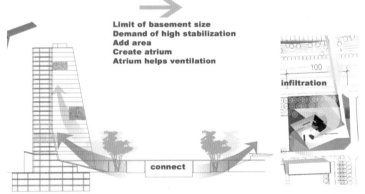

Limit of basement size
Demand of high stabilization
Add area
Create atrium
Atrium helps ventilation

connect

infiltration

通风示意 VENTILATION

**Facade detail** 立面细部

**Chinese traditional sun shading** 中国传统遮阳方式

**Application in our design** 在

2003年/中国北京/高层生态写字楼 | 163

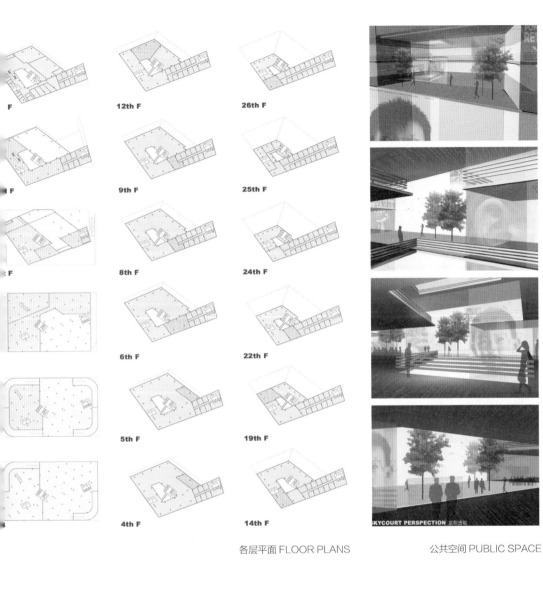

各层平面 FLOOR PLANS　　　公共空间 PUBLIC SPACE

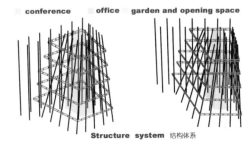

Structure system 结构体系　　　细部设计 DETAIL DESIGN

# E 空间 / E-SPACE

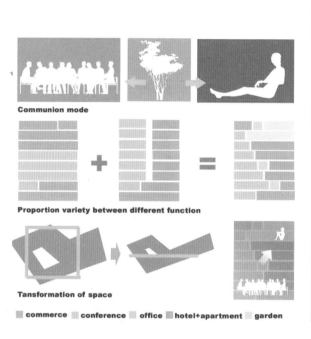

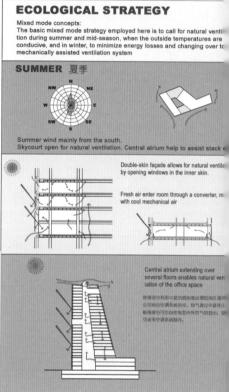

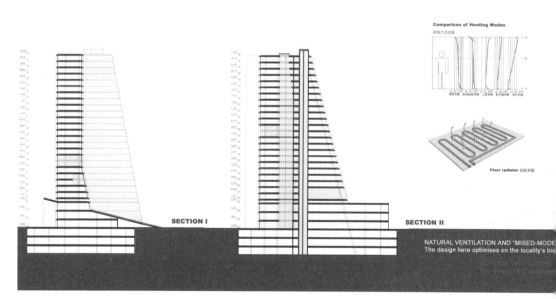

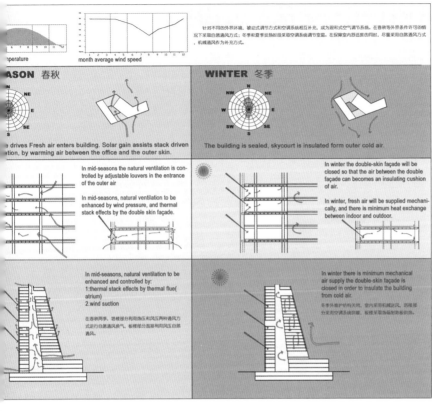

生态策略与环境模拟
ECOLOGICAL STRATEGY AND SIMVLATION

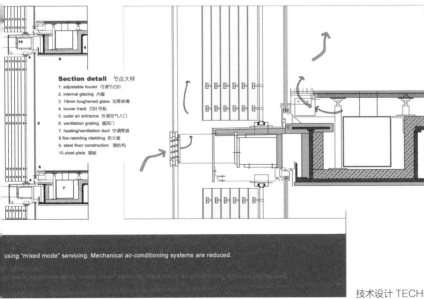

技术设计 TECHNICAL DESIGN

2002年
# 法国索姆 / 滨海艺术家村
## ARTIST'S VILLAGE BY SEA, SOMME, FRANCE

指导教师 栗德祥 邹欢 / 学生 陈帆 高岩 权虹 张颖 陈珊
Teacher: Li Dexiang, Zou Huan
Student : Chen Fan, Gao Yan, Quan Hong, Zhang Ying, Chen Shan

2002年/法国索姆/滨海艺术家村 | 167

2002年的设计地段位于法国索姆省诺曼底海滩，乌尔代勒河入海处。大西洋的潮汐给地段带来了风云变幻的神秘特点，奇妙的光线，空旷无人的海滩，最适合艺术家进行创作和生活。学生们面对的挑战是如何在这样一个没有任何建成环境的地段设计适合艺术家的住宅，需要综合考虑环境、文化、技术以及生活本身的种种要求。得奖方案"Murmur"借用法语的词汇游戏，在建筑群体形态上创造出"墙"这样一个保护生活的形式含义，同时又营造出"呢喃私语"这样一个具有鲜明个性的空间氛围，这也是中国学生对法国文化的独特理解。

In 2002, the site was located at Normandy coast, Somme, France, where the River Urdel enters the sea. The Atlantic tide endows the site with changing mystical features. The wonderful rays of light and empty beach are most favorable for creation and work of artists. The challenge facing students was how to design an artist residence on such a site without any established construction. Therefore, students needed to comprehensively consider the requirements for environment, culture, technique and life. The winning project "Murmur", inspired from a French word, realised a physical symbol of "walls" for protecting life and also created a "murmuring" atmosphere of space with distinctive character. This was a unique comprehension of Chinese students to the French culture.

参与院校
清华大学（中国）
巴黎拉维莱特国立高等建筑学院（法国）
米兰理工大学（意大利）
汉阳大学（韩国）
那不勒斯大学（意大利）

获一等奖院校: 巴黎拉维莱特国立高等建筑学院
清华获奖情况: 陈帆、高岩、权虹、张颖、
　　　　　　　陈珊荣获优秀奖

# 墙 / MURmur

指导教师 栗德祥 邹欢 / 学生 陈帆 高岩 权虹 张颖 陈珊 / 2002 年
Teacher: Li Dexiang, Zou Huan / Student: Chen Fan, Gao Yan, Quan Hong, Zhang Ying, Chen Shan / 2

**教师评语**

方案以墙为基本概念，在总图布局和单体平面中用不同的手段直接体现或间接隐喻"墙"—这个建筑的基本元素，富有艺术与哲学的想象，符合艺术家村和地段与世隔绝的特殊气质。

**COMMENTS**

Taking walls as the basic concept, this project gives a direct manifestation or an indirect metaphor to the "wall" - one of the basic elements of a building, by different approaches in the master plan and individual design. This is abundant in artistic and philosophical imagination and also conforms to the isolated and specialized atmosphere of an artist village and the site.

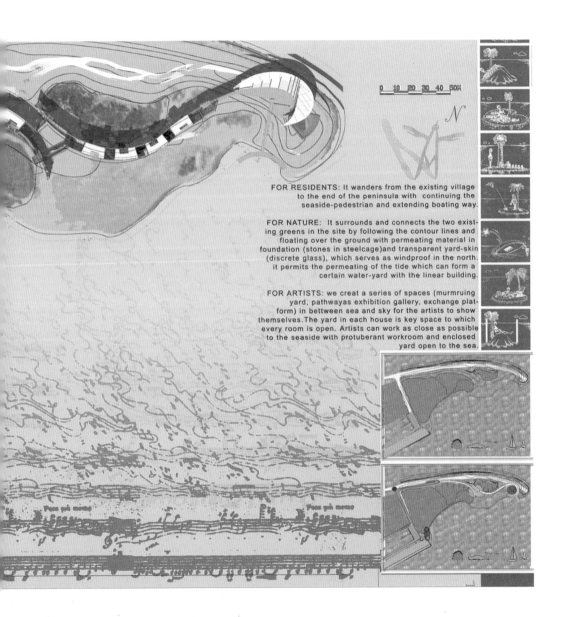

# 墙 /MURmur

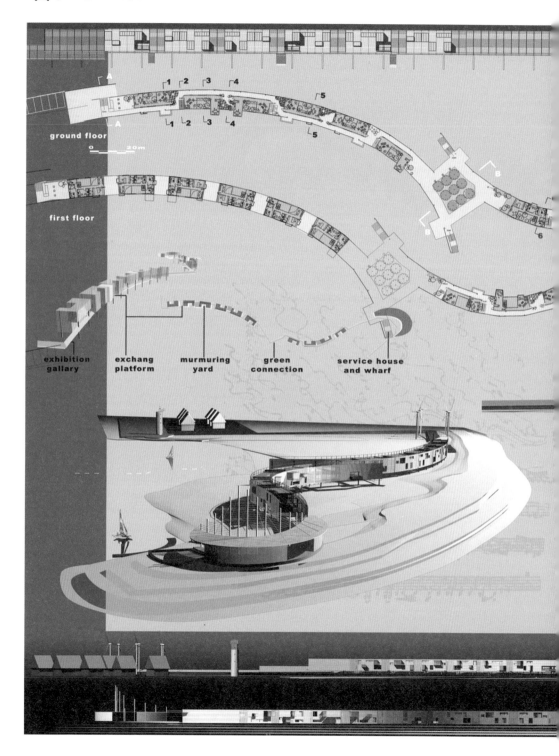

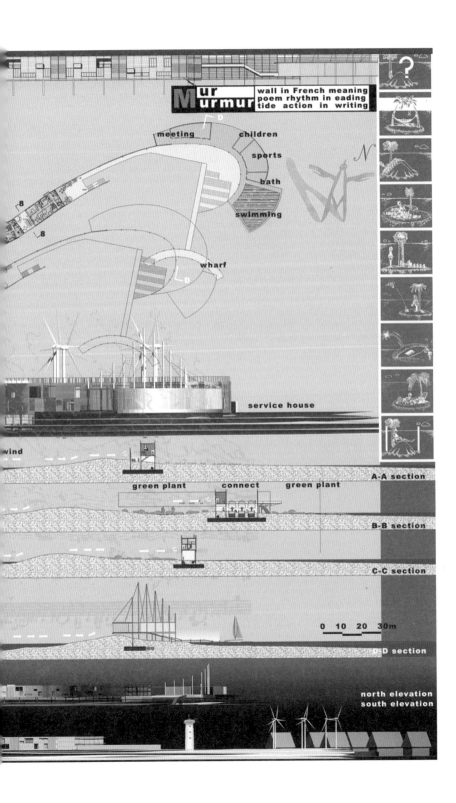

# 墙 /MURmur

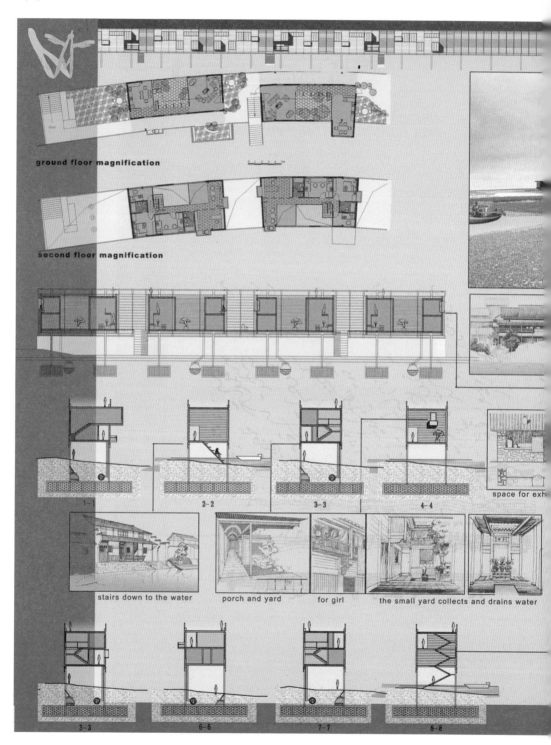

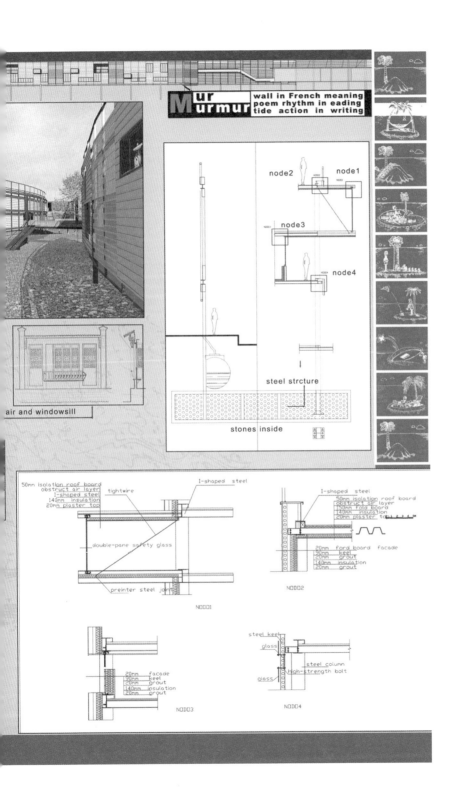

# 附录1 国际建筑工坊历年教学统计 2002–2015
## TEACHING STATISTIC OF AIAC 2002-2015

| 年份 | 主办国家/城市 | 设计题目 | 清华指导教师 | 清华参与学生 | 参与院校 | 评图地点 | 一等奖院校 | 清华获奖情况 |
|---|---|---|---|---|---|---|---|---|
| 2002 | 法国/索姆 | 滨海艺术家村 | 栗德祥 邹欢 | 陈帆、高岩、权虹、张颖、陈珊 | | | 巴黎拉维莱特国立高等建筑学院 | 陈帆、高岩、权虹、张颖、陈珊荣获优秀奖 |
| 2003 | 中国/北京 | 生态高层综合体 | 栗德祥 程晓青 | 庞聪、储以平、李楠、陈晶、张维、田宏、卜骁骏、解霖、赵扬 | 清华大学、巴黎拉维莱特国立高等建筑学院、米兰理工大学、汉阳大学、那不勒斯大学 | 意大利/莱柯 | 米兰理工大学 | 储以平、庞聪、李楠荣获优秀奖 |
| 2004 | 韩国/首尔 | 现代艺术博物馆 | | 雷亮、申吉秀(韩国)、郑天、初腾飞、喻晓、高博、李乃桢、余知衡、霍顺利、赖延璨(中国台湾)、李海燕、海琳娜·毕玛丽(德国)、白洁、陈瑾羲、吴懿、郑姗姗、邓建、杨晓昕、滕静茹、郭佳 | | | 汉阳大学 | 雷亮、申吉秀、郑天荣获优秀奖 |
| 2005 | 意大利/巴勒莫 | 旧港区更新计划 | | 霍振舟、安德雷(哥伦比亚)、梁思愚、刘聪、许阳、范小棣、张阳、张婷 | 清华大学、巴黎拉维莱特国立高等建筑学院、米兰理工大学、汉阳大学、那不勒斯大学、巴勒莫大学 | 法国/巴黎 | 米兰理工大学 | 霍振舟荣获优秀奖 |
| 2006 | 法国/马赛 | 军事遗迹改造与再利用 | 栗德祥 程晓青 邹欢 | 兰俊、李德熙(韩国)、王静、陈洁、丁侬菲、高珊、葛鑫、黄俊杰 | 清华大学、巴黎拉维莱特国立高等建筑学院、米兰理工大学、汉阳大学、那不勒斯大学、巴勒莫大学、达姆施达德工业大学 | 中国/北京 | 巴黎拉维莱特国立高等建筑学院 | 兰俊荣获二等奖 李德熙荣获三等奖 王静荣获优秀奖 |
| 2007 | 马来西亚/马六甲 | 红树林湿地生态旅游中心 | | 张烨、李旭、国萃、黄愈轩、车京恩(韩国)、黄怀海、刘芳、睢蔚、刘晓霖、全成浩(韩国) | 清华大学、巴黎拉维莱特国立高等建筑学院、米兰理工大学、汉阳大学、巴勒莫大学、马来西亚理工大学 | 意大利/西西里 | 清华大学 | 张烨、李旭荣获一等奖 国萃、黄愈轩荣获二等奖 |
| 2008 | 中国/深圳 | 蛇口区滨海创意中心 | | 熊星、李烨、陈忱、秦嘉煜、乔会卿、都文娟、郑娟、孔媛、林展鹏(中国香港)、吴焕、马晓英、傅平川、赵鹏 | 清华大学、巴黎拉维莱特国立高等建筑学院、汉阳大学、庆尚大学、深圳大学 | 法国/巴黎 | 清华大学 | 熊星、秦嘉煜荣获一等奖 李烨、陈忱荣获二等奖 |
| 2009 | 韩国/晋州 | 泗川滨海现代音乐厅 | | 戴天行、阎晋波、曲直、金世中(韩国)、苏海星、赵瞳、孙小暖、揭小凤 | 清华大学、巴黎拉维莱特国立高等建筑学院、汉阳大学、庆尚大学、罗马大学 | 意大利/罗马 | — | — |

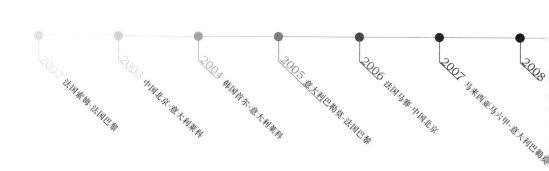

附录1 "国际建筑工坊" 历年教学统计

| 年份 | 主办国家/城市 | 设计题目 | 清华指导教师 | 清华参与学生 | 参与院校 | 评图地点 | 一等奖院校 | 清华获奖情况 |
|---|---|---|---|---|---|---|---|---|
| 2010 | 法国/贝约纳 | 滨水空间更新计划 | 程晓青 邹欢 | 李昆、齐际、刘伯宇、秦岭、吴锡嘉、任凯、王禹、唐任杰 | | 中国/北京 | 清华大学 | 李昆、齐际荣获一等奖 刘伯宇、秦岭荣获优秀奖 |
| 2011 | 法国/奥什 | 军事遗迹改造与再利用 | | 朱源、顾志琦、刘栎珣、于梦瑶、康茜、符传庆、吴艳珊、吴家东、王舒轶、王焓 | 清华大学、巴黎拉维莱特国立高等建筑学院、汉阳大学、庆尚大学、哈尔滨工业大学、沈阳建筑大学 | 韩国/晋州 | 清华大学 | 朱源、顾志琦、刘栎珣、于梦瑶、康茜、符传庆荣获总图设计一等奖 朱源荣获单体设计一等奖 |
| 2012 | 意大利/威尼斯 | 大运河滨水建筑更新设计 | | 陶一兰、钟琳、刘芸、孟宁、孟瑶磊、徐妍、缪一新 | 清华大学、巴黎拉维莱特国立高等建筑学院、汉阳大学、庆尚大学、威尼斯建筑大学、哈尔滨工业大学、沈阳建筑大学 | 中国/沈阳 | 清华大学 | 陶一兰荣获一等奖 钟琳、刘芸、孟宁荣获优秀奖 |
| 2013 | 日本/东京 | 日本桥地区城市更新计划 | | 张冰洁、连晓刚、周南、李明扬、邓施莹、王若凡、李佳婧、茹笑岚、龚梦雅 | 清华大学、巴黎拉维莱特国立高等建筑学院、汉阳大学、庆尚大学、威尼斯建筑大学、哈尔滨工业大学、庆应义塾 | 法国/巴黎 | 巴黎拉维莱特国立高等建筑学院 | 张冰洁荣获二等奖 连晓刚荣获优秀奖 |
| 2014 | 哥伦比亚/波哥大 | 安第斯大学国际学生中心 | | 闵嘉剑、熊哲昆、张华西、李金泰 | 清华大学、巴黎拉维莱特国立高等建筑学院、汉阳大学、庆尚大学、威尼斯建筑大学、庆应义塾、安第斯大学、利马大学 | 意大利/威尼斯 | 威尼斯建筑大学 | 闵嘉剑荣获优秀奖 |
| 2015 | 越南/河内 | 东兴市场区域城市更新计划 | | 李冬、张晨阳、曾睿之、厉之昀、陈茜、甘佩佩（印度尼西亚） | 清华大学、巴黎拉维莱特国立高等建筑学院、汉阳大学、庆尚大学、威尼斯建筑大学、庆应义塾、马德里理工大学、越南国家工程技术大学、广州美术学院、马来西亚理工大学、帕纳空皇家师范大学 | 西班牙/马德里 | 越南国家工程技术大学 | 李冬荣获二等奖 |

2009 韩国晋州-意大利罗马　2010 法国巴约纳-中国北京　2011 法国奥什-韩国晋州　2012 意大利威尼斯-中国沈阳　2013 日本东京-法国巴黎　2014 哥伦比亚波哥大-意大利威尼斯　2015 越南河内-西班牙马德里

# 附录 2　国际建筑工坊参与院校统计 2002-2015
## STATISTIC OF AIAC 2002-2015

**亚洲院校**

中国（5所）：清华大学、哈尔滨工业大学、沈阳建筑大学、深圳大学、广州美术学院
韩国（2所）：汉阳大学、庆尚大学
日本（1所）：庆应义塾
马来西亚（1所）：马来西亚理工大学
越南（1所）：越南国家工程技术大学
泰国（1所）：帕纳空皇家师范大学

**欧洲院校**

法国（1所）：巴黎拉维莱特国立高等建筑学院
意大利（5所）：米兰理工大学、威尼斯建筑大学、那不勒斯大学、巴勒莫大学、罗马大学
德国（1所）：达姆施达德工业大学
西班牙（1所）：马德里理工大学

**南美洲院校**

哥伦比亚（1所）：安第斯大学
秘鲁（1所）：利马大学

# 致谢

首先,感谢法国巴黎拉维莱特国立高等建筑学院 ERIC DUBOSC 教授和意大利米兰理工大学建筑学院 ETTORE ZAMBELLI 教授,多年前在你们的高瞻远瞩和积极推动之下,"国际建筑工坊"得以诞生,一份来自巴黎的美好邀约使清华和"国际建筑工坊"结下不解之缘。感谢十四年来陪伴我们一起成长的各国院校和同行们,在长期的交流合作中我们相互学习、彼此激励、获益匪浅。感谢和"国际建筑工坊"一起度过的那 5000 多个日日夜夜,我们收获了人生中许多难忘的时光。

同样,感谢参与"国际建筑工坊"课程设计研究的 100 多位清华同学,和你们一起分享的悲伤和喜悦仿佛仍旧历历在目,你们的热情和努力为清华赢得了崇高的荣誉。感谢为本课程教学提供鼎力支持的诸位评图老师,你们渊博的学识和清晰的讲评为我们拓展了广阔的视野。感谢给予本课程大力支持的清华大学建筑学院和深圳招商房地产有限公司,你们的帮助令我们始终无畏前行。

最后,感谢为本书作品整理和排版设计做出辛勤工作的王若凡、张华西和唐丽等同学,感谢清华大学出版社的张占奎、周莉桦诸位编辑,在你们的帮助之下我们的成果才得以呈现。本书并不仅仅是一份教学成果的记录,更为重要的是它记录了各国建筑学人探索国际教育合作的心得与感悟,相信我们的共同努力一定会为建筑教育事业的发展做出积极贡献。

栗德祥　　　　　　　程晓青　　　　　　　邹欢